Chemical Principles of Environmental Pollution

B.J. Alloway
University of Reading
and
D.C. Ayres
Queen Mary and Westfield College
University of London

QMW LIBRARY
(MILE END)

BLACKIE ACADEMIC & PROFESSIONAL
An Imprint of Chapman & Hall

London · Glasgow · New York · Tokyo · Melbourne · Madras

Published by
Blackie Academic & Professional, an imprint of Chapman & Hall,
Wester Cleddens Road, Bishopbriggs, Glasgow G64 2NZ

Chapman & Hall, 2–6 Boundary Row, London SE1 8HN, UK

Blackie Academic & Professional, Wester Cleddens Road, Bishopbriggs, Glasgow G64 2NZ, UK

Chapman & Hall Inc., 29 West 35th Street, New York NY10001, USA

Chapman & Hall Japan, Thomson Publishing Japan, Hirakawacho Nemoto Building, 6F, 1–7–11 Hirakawa-cho, Chiyoda-ku, Tokyo 102, Japan

DA Book (Aust.) Pty Ltd., 648 Whitehorse Road, Mitcham 3132, Victoria, Australia

Chapman & Hall India, R. Seshadri, 32 Second Main Road, CIT East, Madras 600 035, India

First edition 1993

© 1993 Chapman & Hall

Typeset in 10/12 pt Times New Roman by Photoprint, Torquay
Printed in Great Britain by the Alden Press, Oxford

ISBN 0 7514 0013 0

Apart from any fair dealing for the purposes of research or private study, or criticism or review, as permitted under the UK Copyright Designs and Patents Act, 1988, this publication may not be reproduced, stored, or transmitted, in any form or by any means, without the prior permission in writing of the publishers, or in the case of reprographic reproduction only in accordance with the terms of the licences issued by the Copyright Licensing Agency in the UK, or in accordance with the terms of licences issued by the appropriate Reproduction Rights Organization outside the UK. Enquiries concerning reproduction outside the terms stated here should be sent to the publishers at the Glasgow address printed on this page.
    The publisher makes no representation, express or implied, with regard to the accuracy of the information contained in this book and cannot accept any legal responsibility or liability for any errors or omissions that may be made.

A catalogue record for this book is available from the British Library

Library of Congress Cataloging-in-Publication data

Alloway, B. J.
    Chemical principles of environmental pollution / by B.J. Alloway and D.C. Ayres. -- 1st ed.
        p.   cm.
    Includes bibliographical references and index.
    ISBN 0–7514–0013–0 (pbk.)
    1. Pollution. 2. Environmental chemistry.   I. Ayres, D. C.
II. Title.
TD193.A45 1993
628.5--dc20   93–3428
                                                                  CIP

Preface

Pollution is the most serious of all environmental problems and poses a major threat to the health and well-being of millions of people and global ecosystems. Other major environmental problems are either wholly or partly caused by pollution; these include global warming, climatic change and the loss of biodiversity through the extinction of many species. In the future, environmental pressures can only increase as a result of population growth and the expectation of higher living standards.

For at least thirty years people have become increasingly aware of these issues. As a result governments and regulatory bodies have responded by taking action against grossly polluting activities and by enforcing tighter limits on the emission of pollutants into the environment. As the level of control improves so the financial costs increase exponentially, hence an effective limit must be imposed which does not impose unacceptable burdens on industrial producers.

As a consequence of the increasing economic and legal pressures which result from pollution control, there is a growing need for greater understanding of the scientific principles underlying environmental pollution. Appropriately qualified professionals will be needed in the energy, manufacturing, service and waste disposal industries and their regulatory authorities. They will be required to enforce increasingly strict standards and to monitor the environment for accidental pollution. The rapid growth in the numbers studying for undergraduate and postgraduate degrees and related qualifications in environmental science is a reflection of these requirements. Modules dealing with aspects of pollution have been introduced into many courses in the traditional disciplines of chemistry, biology, geography and civil engineering.

This book was written, in sympathy with these trends, for students and others with a basic knowledge of chemistry. It provides an introduction to the principles, and adopts a pollutant-orientated approach, rather than the more common one based on particular sectors such as air or water. All the main groups of substances are covered, and the principles relating to their nature, sources, transport, environmental behaviour and effects on targets. It is expected that readers will want to consult more advanced specialized texts dealing with particular topics and an extensive bibliography has been included to further their studies.

The authors have been involved in teaching and researching this subject for many years; both have been Directors of the B.Sc. Programme in Environmental Science which began at Westfield College and now continues in the merged Queen Mary and Westfield College of the University of London. We originally came to the subject from different backgrounds (BJA from soil science and DCA from organic chemistry) but, as is inevitable in this research, we both now have an interdisiplinary approach. Nevertheless, it must be stressed that the study and management of environmental pollution requires the rigorous application of related scientific principles. Interdisciplinary environmental science is not 'soft' science; it is intellectually demanding, rewarding and urgently needed in our polluted world.

BJA
DCA

Contents

PART ONE   BASIC PRINCIPLES

1 Introduction   **3**

    1.1 Pollution in the modern world   4
    1.2 Definition of pollution   5
    References   13
    Further reading   13

2 Transport and behaviour of pollutants in the environment   **16**

    2.1 A basic model of environmental pollution   16
    2.2 Sources of pollutants   17
    2.3 The pollutants   17
        2.3.1   Classification of hazardous substances in the USA   17
        2.3.2   European Community Dangerous Substances Directive   19
        2.3.3   UK priority list of pollutants   21
        2.3.4   Pesticides   21
        2.3.5   Indoor pollution   23
    2.4 Physical processes of pollutant transport and dispersion   24
        2.4.1   Transport media   24
        2.4.2   Transport of pollutants in air   24
        2.4.3   Some important types of reactions which pollutants
                undergo in the atmosphere   31
    2.5 Transport of pollutants in water   31
        2.5.1   Biochemical processes in water (involving microorganisms)   33
    2.6 The behaviour of pollutants in the soil   35
        2.6.1   The composition and physico-chemical properties of soils   35
        2.6.2   Cation and anion adsorption in soils   37
        2.6.3   Adsorption and decomposition of organic pollutants   39
    2.7 Concluding remarks   42
    References   42

3 Toxicity and risk assessment of environmental pollutants   **44**

    3.1 Basic principles of toxicology   44
    3.2 Effects of pollutants on animals and plants   47
        3.2.1   Effects of pollutants on humans and other mammals   47
        3.2.2   Teratogenesis, mutagenesis, carcinogenesis and immune
                system defects   48
        3.2.3   Ecotoxicology   50
    3.3 Assessment of toxicity risks   51
        3.3.1   Pollutants in contaminated land   51
        3.3.2   Pollutants in drinking water   56
        3.3.3   Toxic or explosive gases and vapours   57
    References   58

**4 Analysis and monitoring of pollutants – organic
compounds**   **59**

4.1 Chromatography   59
4.2 Thin layer chromatography (TLC)   60
    4.2.1 Separation of pesticides   61
    4.2.2 Separation of metal cations   61
4.3 Gas liquid chromatography (GLC)   62
    4.3.1 Detection of eluted substances   65
    4.3.2 Principal parameters   67
    4.3.3 Optimum operating conditions   68
    4.3.4 Capillary columns for GLC   70
    4.3.5 Analysis of urban air pollution   71
    4.3.6 Detection by mass spectrometry   71
4.4 High pressure liquid chromatography (HPLC)   77
    4.4.1 The components   78
    4.4.2 Detectors   80
    4.4.3 Analysis of polluted air   81
    4.4.4 Analysis of polluted water   85
    4.4.5 Trace enrichment followed by GLC analysis   86
4.5 Pollution by metals – atomic absorption spectroscopy   86
    4.5.1 Historical   86
    4.5.2 Basic theory of atomic absorption and emission   88
    4.5.3 The Lambert–Beer law   90
    4.5.4 Instrumental details   91
    4.5.5 Interferences   92
    4.5.6 The determination of sodium in concrete by AAS   93
    4.5.7 Sample preparation   94
    4.5.8 Precision and accuracy of measurement   94
    4.5.9 Graphite furnace AAS   95
4.6 A plasma source   95
    4.6.1 ICP-mass spectrometry   97
4.7 Analytical quality assurance   97
4.8 Environmental monitoring   99
    4.8.1 Introduction   99
    4.8.2 Monitoring emissions   100
References   104
Further reading   105

PART TWO   THE POLLUTANTS

5 Inorganic pollutants   **109**

5.1 Ozone   109
    5.1.1 Historical   109
    5.1.2 Formation   109
    5.1.3 Physical properties and structure   109
    5.1.4 The ozone layer   110
    5.1.5 Factors which disturb the natural environment   111
    5.1.6 Chemistry of stratospheric CFC   112
    5.1.7 Control measures   113
    5.1.8 Ozone in the troposphere   114
    5.1.9 Diurnal variations of ozone levels   115
    5.1.10 Toxicity and control   115
5.2 Oxides of carbon, nitrogen and sulphur   115
    5.2.1 Carbon dioxide   116
    5.2.2 Oxides of nitrogen   124
    5.2.3 Oxides of sulphur   129
5.3 Heavy metals   140
    5.3.1 General properties   140

	5.3.2	Biochemical properties of heavy metals	141
	5.3.3	Sources of heavy metals	142
	5.3.4	Environmental media affected	149
	5.3.5	Heavy metal behaviour in the environment	151
	5.3.6	Toxic effects of heavy metals	155
	5.3.7	Analytical methods	158
	5.3.8	Examples of specific heavy metals	159
5.4	Other metals and inorganic pollutants	164
	5.4.1	Aluminium	164
	5.4.2	Beryllium	165
	5.4.3	Fluorine	166
5.5	Radionuclides	167
	5.5.1	History and nomenclature	167
	5.5.2	Types of radioactive emission	168
	5.5.3	Units of energy and measurement of toxicity	169
	5.5.4	Radioactive potassium	170
	5.5.5	Production of radionuclides by artificial means	171
	5.5.6	Nuclear fission	171
	5.5.7	Power generation in nuclear reactors	173
	5.5.8	Nuclear reactor types	174
	5.5.9	The future of nuclear power	178
	5.5.10	Observations on major accidents	180
	5.5.11	Radioactive release within buildings	184
	5.5.12	Social aspects of nuclear power generation	186
	5.5.13	Power from thermal fusion	186
	5.5.14	Cold fusion	188
5.6	Mineral fibres and particles	188
	5.6.1	General aspects	188
	5.6.2	Analysis	189
	5.6.3	Examples of mineral pollutants	189
	References	191
	Further reading	194

6 Organic pollutants   **196**

6.1	Smoke	196
6.2	Methane and other hydrocarbons – coal and oil as sources	199
	6.2.1	The formation of coal	199
	6.2.2	Petroleum	200
	6.2.3	Methane	202
	6.2.4	Higher alkanes	203
	6.2.5	Polycyclic aromatic hydrocarbons (PAH)	206
6.3	Organic solvents	210
	6.3.1	Adhesives	211
	6.3.2	Coatings and inks	212
	6.3.3	Aerosol sprays	213
	6.3.4	Metal cleaning	214
	6.3.5	Dry cleaning of clothes	214
	6.3.6	Solvent toxicology	215
	6.3.7	Organochlorine compounds	216
	6.3.8	Detergents	217
	6.3.9	Indoor pollution	220
6.4	Organohalides: pesticides, PCBs and dioxins	221
	6.4.1	Historical	221
	6.4.2	Organochlorine production	222
	6.4.3	DDT (dichlorodiphenyl trichloroethane)	222
	6.4.4	Lindane, hexachlorocyclohexane	223
	6.4.5	Some other chlorinated pesticides	225
	6.4.6	Organochlorine herbicides	226
	6.4.7	Toxic effects of insecticides	227

	6.4.8	Control of pesticides	229
	6.4.9	Vinyl chloride and polyvinyl chloride	231
	6.4.10	Polychlorobiphenyls	232
	6.4.11	Toxic substances in herbicides	235
	6.4.12	Metabolism of chloraromatic compounds	242
	6.4.13	Disposal of organochlorine compounds	242
	6.4.14	Cremation or burial	243
	6.4.15	Use of decay organisms	244
6.5	Natural, organophosphorus and carbamate pesticides		244
	6.5.1	Naturally occurring pesticides	244
	6.5.2	Organophosphorus pesticides	248
	6.5.3	Carbamate pesticides	253
6.6	Odours		255
	6.6.1	Important properties of odours	255
	6.6.2	Methods of odour control	256
	6.6.3	Methods of odour treatment	256
	References		257
	Further reading		259

PART THREE   WASTES AND OTHER MULTI-POLLUTANT SITUATIONS

7   Wastes and their disposal   **263**

7.1	Introduction	263
7.2	Amounts of waste produced	263
	7.2.1 Industrial wastes	264
	7.2.2 Municipal wastes	264
7.3	Methods of disposal of municipal wastes	264
	7.3.1 Landfilling	264
	7.3.2 Incineration	265
	7.3.3 Composting	266
	7.3.4 Recycling	266
7.4	Sewage treatment	267
7.5	Hazardous wastes	270
	7.5.1 The nature and amount of hazardous waste produced	270
	7.5.2 Hazardous waste management	271
	7.5.3 New technologies for waste disposal	273
7.6	Long-term pollution problems of abandoned landfills containing hazardous wastes	274
	7.6.1 Love Canal, New York, USA	274
	7.6.2 Lekkerkirk, near Rotterdam, The Netherlands	275
7.7	Tanker accidents and oil spillages at sea	275
7.8	Other multi-pollutant situations	277
7.9	Chemical time bombs	278
	References	278

Appendix   Table of units and conversions   **280**

Index   **281**

Part One

Basic Principles

KU-318-829

Introduction 1

The dramatic increase in public awareness and concern about the state of the global and local environments which has occurred in recent decades has been accompanied and partly prompted by an ever growing body of evidence on the extent to which pollution has caused severe environmental degradation. The introduction of harmful substances into the environment has been shown to have many adverse effects on human health, agricultural productivity and natural ecosystems. However, it is increasingly surprising just how resilient global environmental systems are to many of the pollutant burdens imposed upon them. Nevertheless, the instances where 'chemical time bombs' have had dramatic ecological effects, such as in the forests of Central Europe where many years of inputs of SO_2 and other atmospheric pollutants eventually led to a widespread die-back in conifers (see section 5.2), should remind us that we cannot be complacent about environmental pollution. It is therefore very important for as many people as possible to appreciate the extent of pollution, its causes, the substances involved, their biological and environmental effects, and methods of controlling and rectifying pollution.

This book introduces the reader to the basic principles relating to the main types of environmental pollutants, their sources, chemical properties and the reactions they undergo in the air, water and soil. It is felt that the use of a pollutant orientated approach is more logical and appropriate for an introductory text for science students than a more ecologically focused or environmental media approach. The effects of pollutants on organisms and the environment are best studied after the nature of the chemicals involved and some basics about their environmental behaviour are understood. However, the subjects of transport and behaviour of pollutants in the environment, principles of toxicology, assessment of the risks which pollutants pose and the methods of monitoring and analysis are briefly covered on a general basis at the beginning of the book and referred to later in more detail in the sections dealing with the individual groups of pollutants. The last section deals with wastes and their disposal and multi-pollutant situations. In a book of this type it is not possible to give an exhaustive coverage of every important pollutant, nor to discuss all relevant aspects of environmental behaviour in detail. It is expected that the reader will want to follow up this introduction by referring to more specialized texts and some suggested titles for further reading are provided. In many cases the books that give a more detailed coverage tend to focus on one particular medium, such as air or water, or a restricted group of pollutants, such as heavy metals or pesticides.

1.1
Pollution in the modern world

Although problems such as the destruction of valued environments, soil erosion and the extinction of species are very important for the future of mankind, it is pollution which arouses the most interest. This is because people realize that pollution impacts on them directly through effects on their health, their food supply, the degradation of buildings, and other items of cultural heritage, as well as overt effects on forests, rivers, coastlines and ecosystems that they are familiar with. The costs of these effects in the depreciation of resources, lost productivity and in cleaning up or improving polluted environments are high and are increasingly occupying the attention of governments and politicians around the world, especially in technologically advanced countries. With increased legislation intended to reduce and prevent pollution there is a corresponding increase in the involvement of specialist consultants, lawyers, insurers and financiers. There are many members of a wide range of professions with a growing interest in environmental pollution.

Although cases of overt pollution have occurred since the beginning of the Industrial Revolution, the far greater awareness about pollution among the populations of most advanced countries is largely due to the mass media (especially television and radio) which enables millions of people to know when a major environmental catastrophe, for example the wreck of the oil tanker *Brear* in the Shetland Isles, is happening. The spectacular TV images and radio reports of the massive oil slick caused by the spillage of 85 000 t of crude oil from the floundering oil tanker and its toll of dead and dying sea birds and mammals make most people aware, for a short time at least, of the fragility of the environment (see section 7.7). However, what is less generally appreciated, but much more important overall, is that most environmental pollution is insidious and its harmful effects only become apparent after long periods of exposure. Many people are exposed to pollutants, which may cause cancer ten or twenty years later, without realizing it (see chapter 3). Gradual increases in atmospheric pollution can be causing chronic toxic effects in trees which do not appear for 20 or more years and are irreversible. Likewise, lakes can become increasingly polluted (see section 5.2.3) and species die out without many obvious signs, at least in the early stages. For this reason, environmental monitoring has become recognized as being vitally important in detecting where insidious pollution is occurring, the pollutants involved and the sources from which they came. Environmental monitoring has benefited from the development over recent decades of rapid and accurate methods of chemical analysis, such as gas and liquid chromatography for organic pollutants and atomic absorption spectrophotometry and inductively-coupled plasma atomic emission spectrometry for metals (see chapter 4).

It is estimated that over the whole period of human history, around 6×10^6 chemical compounds have been created, most of these within this century. Nowadays, about 1000 new compounds are being synthesized each year and between 60 000 and 95 000 chemicals are in current commercial use. Pollutant chemicals are responsible for many human illnesses, including chronic bronchitis associated with air pollution, especially SO_2, and neurological conditions linked to both Hg and Pb pollution. The latter pollutant is so ubiquitous through its use in petrol for motor vehicles and its presence in the smoke from coal combustion that there are few parts of the world not affected by it, at least to a slight extent. In fact it can be argued that all soils in industrialized countries have been polluted with many trace substances including Cd, Pb, polycyclic aromatic hydrocarbons (PAHs), polychlorinated biphenyls (PCBs) and polychlorinated dibenzo-p-dioxins (PCDDs) and furans (PCDFs) by aerial and other inputs (Jones, 1991) (see sections 5.3, 6.2.5 and 6.4).

The incidence of asthma and 'sick office' syndrome and other possible respiratory allergies are considered to be linked to people's exposure to an ever increasing range of chemicals in the atmosphere, especially in urban and industrial areas. It has recently become apparent that there is a significant decrease in the production of viable human sperm in technologically advanced countries and this is also thought to be due, at least in part, to exposure to pollutants. Many toxic chemicals are used regularly in our daily lives because their particular use is not thought to constitute a hazard to health and the advantages of using them outweigh their possible risks. An example of this is the use of chlorine compounds for water disinfection and other hygiene purposes. In many Third World countries the lack of a hygienic water supply is the cause of the deaths of many young children. Safe drinking water is taken for granted in most developed countries and this usually necessitates disinfection with Cl_2, especially when water is recycled several times. However, the formation of trace chlorinated organic compounds can occur in treated waters and there appears to be a weak correlation between chlorinated water and bladder cancer. On balance, the current view is that the dangers from non-disinfected water outweigh the slight risk of cancer (see chapter 3).

1.2
Definition of pollution

A widely used definition of pollution is 'the introduction by man into the environment of substances or energy liable to cause hazards to human health, harm to living resources and ecological systems, damage to structures or amenity, or interference with legitimate uses of the environment' (Holdgate, 1979). Some experts make a distinction between contamination and pollution. Contamination is used for situations where a substance is present in the environment, but not causing any obvious harm, while pollution is reserved for cases where harmful effects are apparent. However, the problem with this distinction is that with improved methods of analysis and diagnosis, it may become apparent that harmful effects have been caused and so situations initially described as contamination may have really been pollution. Holdgate's definition avoids this problem.

Pollutants are basically of two types: *primary pollutants*, which exert harmful effects in the form in which they enter the environment, and *secondary pollutants* which are synthesized as a result of chemical processes, often from less harmful precursors, in the environment. Although highly toxic substances are responsible for many cases of environmental pollution, under some circumstances materials which are normally considered harmless may cause pollution if they are present in excessive quantities or in the wrong place at the wrong time, and this makes definitions difficult (BMA, 1991). For example, milk or sugar are not normally considered as environmental pollutants. However, if a lorry carrying a load of either of these important foodstuffs was to spill much of its load into a river, severe pollution due to the high biochemical oxygen demand (BOD) of these substances would result in the death of many fish (see chapter 3).

In all cases of pollution there is (i) a source of pollutants, (ii) the pollutants themselves, (iii) the transport medium (air, water or direct dumping onto land), and (iv) the target (or receptor) which includes ecosystems, individual organisms (e.g. humans) and structures. Pollution can be classified in several ways according to (i) the source (e.g. agricultural pollution), (ii) the media affected (e.g. air pollution or water pollution) or (iii) by the nature of the pollutant (e.g. heavy metal pollution).

Although undesirable and costly, pollution is an inevitable and necessary part of life for most of the world's population especially in large communities and those relying on technology and mechanized transport. Even in primitive cultures, accumulated human excretory products and smoke from cooking fires cause pollution. Where the volumes and rates of emission and toxicities of pollutants are relatively low, environmental processes can usually degrade or assimilate these excesses to a much greater extent than the more toxic, air, water and land pollutants produced in large quantities in more technologically advanced cultures.

Although industrial pollution in the UK in the nineteenth century was severe enough to necessitate Parliament passing the Alkali Act in 1862 and the Rivers Pollution Act in 1876, it took the notorious London smog in 1952 to bring about a major improvement in urban air quality through the introduction of the Clean Air Act of 1956. Smoke and smogs had generally been accepted as an unavoidable fact of life in towns and cities of the UK and many other industrialized countries. In the UK a high incidence of chronic bronchitis ('the English disease') was associated with this type of air pollution. Unfortunately, the reduction of smoke and SO_2 levels which the Clean Air Act (1956) brought about were soon offset by an increase in CO, NO_x PAH, O_3, PbBrCl and PAN concentrations due to increasing numbers of motor vehicles on the roads and the replacement of coal by oil for heating buildings. The pollution produced by a single car on a journey to work is of little consequence, but when thousands of cars are involved a serious pollution situation develops. In conditions of bright sunlight, such as in Los Angeles, Mexico City, Athens and many other cities, even those in temperate areas such as London in the summer, the primary pollutants of motor vehicle exhausts (smoke, PAH, CO, NO_x and PbBrCl) are supplemented by the secondary pollutants O_3, NO_2, and peroxyacetyl nitrate (PAN) synthesized in photochemical reactions and these give rise to a pungent photochemical smog which irritates the eyes, nose and throat and causes more severe toxic effects in humans, animals and plants than the primary pollutants (see sections 5.2, 5.3, 6.2 and 6.3).

The worsening problems of atmospheric pollution from motor traffic in many of the world's cities have led to the introduction of legislation to try and mitigate or prevent the nuisance. This has included the phasing out of Pb additives in petrol and the introduction of catalytic converters in some countries, such as the USA. However, air quality problems still remain serious in most large cities and there appears no way of preventing them completely unless private cars are banned from city centres and environmentally sound public transport systems are introduced. The expected exhaustion of petroleum reserves in around 50 years' time may help to force the issue. However, if there is a move towards using oil derived from shale rocks as a substitute for petroleum, the air quality problems may be exacerbated because the shale oils contain a wider range of compounds which would create pollutants in exhaust gases. The environmental impact of the exploitation of oil shale deposits around the world, in areas such as the Green River Basin in the USA, would also be much greater than those of petroleum (see sections 5.2 and 6.2).

One of the factors which led to an increase in environmental awareness and concern about pollution around 30 years ago was the publication of certain inspiring books. One of the first was *Silent Spring* by Rachel Carson (1963), which focused attention on insidious pesticide pollution. The title dramatically refers to a scenario of a spring without birdsong due to most birds having been killed by pesticides or their residues. At the time this book was published, developments in plant breeding and pest control had boosted cornucopian optimism in advanced countries. Food production was increasing rapidly along with an exponential rise in the use of pesticides and fertilizers (see section 6.4) and little attention was given to the consequences of their accumulation in the environment, or of the toxicity of their degradation products (see sections 3.2 and 6.4). Carson (1962) was instrumental in drawing attention to the risk of cancer in people exposed to many pesticides especially as consumers. Some authors

Figure 1.1 Opencast lignite mining in northern Bohemia (Czech Republic) and air pollution from lignite burning for electricity generation (photo: B.J. Alloway).

consider this to be a serious misconception and have drawn attention to the fact that many normal constituents of the diet pose a greater risk of cancer than very small traces of pesticide residues in foods (Ames and Gold, 1990). Nevertheless, the fact remains that the smoking of cigarettes is by far the greatest risk of developing cancer.

Silent Spring was soon followed by many books on environmental themes but one book which aroused much interest and controversy was *Limits to Growth* by Meadows *et al*. (1972). It is particularly noteworthy because it was one of the first books to be based on predictions from a mathematical simulation model (World 3). It predicted that exponential increases in human population, in the consumption of finite resources and in pollution would lead to a drastic reduction in the population which the earth would be able to sustain over a 100 year period. One very important prediction from the model was that food production would decrease, partly as a consequence of environmental pollution. Nine million copies of the book were sold and, although it attracted much criticism, it focused attention on the exponential rate of increase in the severity of global environmental problems and on the need for concerted international action to try and slow them down.

In the same year that *Limits to Growth* was published (Meadows *et al*.,

1972), the United Nations organized its first international conference on the environment in Stockholm and also set up the United Nations Environmental Programme (UNEP) with its headquarters in Nairobi. The United States had already established its own Environmental Protection Agency in 1970 and nine other countries had also created agencies or government departments concerned with environmental matters. By the time of the second United Nations Conference on the Environment and Development (the so-called 'Earth Summit') in Rio de Janeiro 20 years later (1992), more than 100 countries had their own environmental agencies or ministries.

Pollution problems rapidly escalate in severity when the rate of pollutant emissions exceeds the capacity of the environment to assimilate them. A vivid example of this can be found in north-west Bohemia (Czech Republic). After the Second World War, increased exploitation of the local brown coal (lignite) deposits for electricity generation (14×10^6 t in 1945 to 100×10^6 t in 1987) resulted in a marked increase in atmospheric SO_x pollution due to the coal's high sulphur content (<15%) and lack of pollution control. Most bituminous coals in use for electricity generation tend to contain 0.5–4% sulphur, so this Czech lignite is a very rich source of the harmful pollutant SO_2 (see section 5.2). Although the increasing air pollution rapidly affected human health, some of its effects on the local environment took longer to become apparent. It took nearly twenty years before large numbers of trees in the forest on the tops of the nearby mountains in the Ertzgebirge range started to 'die-back'. This death of trees ('waldsterben' in German) continued to increase in severity and today most of the mountain tops in this part of the country are without trees and it is estimated that around 100 000 ha of forest are affected in the whole of the Czech Republic. The rate of sulphur deposition in the mountains most affected by the forest die-back reached 15 g $S/m^2/year$ which rates it as one of the most severely polluted places in the world (Moldan and Schnoor, 1992). The phenomenon of die-back in trees due to atmospheric pollution is becoming a major cause for concern in many parts of Europe and North America. Problems with the acidification of lakes and the death of many species of aquatic organisms also occur in many areas affected by die-back in trees.

In general, it is possible to summarize the main factors responsible for pollution and other types of environmental deterioration in any community or society as being due to the combined effects of population, affluence and technology (Meadows *et al.*, 1992).

Impact on environment = Population × Affluence × Technology

Basically, the larger the population, the greater the extent of environmental deterioration due to related needs for food production, living space, waste disposal, communications and so on. The world's population rapidly increased after the Second World War. It more than doubled in the

40 years between 1950 (2516.44 million) and 1990 (5292.20 million). It is expected to reach 5770.29 million by 1995 and 8504.22 million by the year 2025. The average crude birth rate (births/1000 population) for the world in 1970–75 was 31.5: it was 46.6 in Africa, 25 in North and Central America, 33.2 in South America, 34.8 in Asia, 15.7 in Europe, 18.1 in what was the USSR, and 23.9 in Oceania (World Resources Institute, 1992). In all continents the birth rate is expected to have decreased slightly for the period 1990–95 (world 26.4, Africa 43.5, Asia 26.9, South America 26.2 and Oceania 18.6). The trend of decreasing birth rate is accompanied by a trend towards longer life expectancy (for the world in 1970–75 this was 58.5 years, in 1990–95 it was 65.5 years) with a larger number of older people in the population. Even though some European countries, such as the UK, Netherlands and Belgium had relatively low rates of population increase in the period 1985–90 (0.22%, 0.63% and 0.03%, respectively) the UK has a population density of more than 200 persons/km^2 and both Belgium and the Netherlands have densities in excess of 300 persons/km^2. This high density of population creates severe environmental problems with the conflicting demands for land for living space, food production, waste disposal and the other many essential uses of land, and the pollution created by people living and working in relatively close proximity. In contrast, France, Greece, the Republic of Ireland, Portugal and Spain have population densities of 100/km^2 or less and consequently there is less pressure from competing uses of land.

Most of the increase in the world's population is occurring in the less developed countries (LDCs) of the Third World where levels of affluence and technology are generally low. It can be concluded that the environmental impact of these rapidly growing populations is not as great as it would be if the populations were more affluent and dependent on technology. Nevertheless, their dependence on technology is slowly increasing; as high yielding varieties of crops with their requirement for fertilizers, pesticides and, possibly, irrigation are introduced, the contribution of these countries to pollution and other aspects of environmental deterioration are likewise increasing. The pollution problem is compounded by the continued use in Third World countries of chemicals, including pesticides such as DDT, and Pb-containing petrol, which have either been banned or are being phased out on environmental grounds in technologically advanced countries (see sections 6.4 and 5.3).

A community's affluence (i.e. its capital stock per person) determines the amount of materials required to maintain this standard of living and hence the greater the level of affluence, the greater the need for the exploitation of natural resources. This, in turn, causes much pollution, for example in the mining, transport and smelting of metals. Likewise, the fabrication of metal products, their corrosion in use and eventual disposal are all potentially polluting activities.

Technology requires energy and so the higher the level of technology in

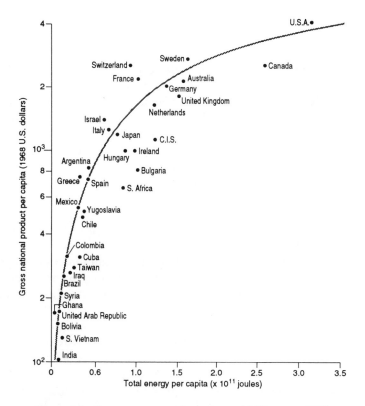

Figure 1.2   Energy use per capita in the world (Simmons, 1987).

a community, the greater the need for energy and the greater the extent of pollution and environmental damage caused by energy generation. This includes the mining of coal, extraction and refining of petroleum, the mining and processing of uranium, the atmospheric and wider environmental pollution from fossil fuel combustion in electricity generation and the disposal of the ash. Nuclear power station accidents, such as that at Chernobyl, nuclear waste reprocessing and disposal are also part of the environmental price that has to be paid for a plentiful supply of convenient energy (see section 5.5). Nevertheless, the relationship between energy and technology can be modified, to a certain extent, by the use of technology to control pollution (e.g. flue gas desulphurization) from energy generation (Figure 1.2).

Energy generation is perhaps the most ubiquitous cause of pollution in the world as a whole, especially if it is taken in its widest context to include internal combustion engines in motor vehicles and fires for cooking and heating. The total world consumption of energy in 1990 was estimated to be 8013×10^6 t of oil equivalent compared with 5171×10^6 t of oil equivalent in 1970, almost a 55% increase over 20 years (Allen, 1992). Although the burning of wood for fuel in many less developed countries will not cause so much environmental pollution as fossil fuels or nuclear power, it still has a marked environmental impact. This fuel wood is, in theory, a renewable energy resource, but in many cases the excessive and often complete removal of wood and other combustible vegetation has led to irreversible land degradation, especially through soil erosion. The use of wood for fuel ranges from 98% of the energy consumed in Nepal, 82% in Nigeria, 63% in Thailand to 33% in Brazil. Wood burning in open fires or primitive stoves results in the production of a large amount of smoke with its associated formation of polycyclic aromatic hydrocarbon (PAH) organo micropollutants which can have deleterious effects on the people constantly inhaling this smoke (see sections 6.1 and 6.2).

In general, cities and urban areas throughout the world with their concentrated population, high consumption of energy, transport and industrial activities, tend to have the worst environmental pollution problems. The disposal of the population's sewage, municipal and industrial wastes and the atmospheric pollution from urban sources impinges on surrounding rural areas. Urban heat island effects tend to confine much of the dust and atmospheric pollutants within the region and this exacerbates the effects of pollution on the population.

The coastlines of densely populated regions are frequently polluted by sewage (see section 7.4). In less developed countries, untreated sewage is frequently discharged into rivers and carried into the sea, but even in more developed countries, sewage from coastal communities is piped directly into the sea. Although the sea disposal of untreated sewage should have been phased out within the European Community by 1998, many tourist beaches currently have poor quality bathing waters due to the presence of bacteria and viruses of sewage origin. Coastal fisheries and shellfish resources have also been significantly affected. In some places, coastal vegetation has been excessively scorched by salt spray due to the presence of detergents of sewage origin in the sea water which have modified the effect of the salt solution droplets on previously resistant plant leaves.

Point source pollution resulting from mining, industrial and military activities occurs all over the world, often in some very remote regions. As a result of atmospheric pollution and the use of fertilizers contaminated with potentially hazardous substances, such as Cd, most of the soils in technically advanced countries are polluted (or contaminated) to a slight extent, although their composition is much less affected than soils near to urban/industrial localities (see section 5.3).

In most cases, air pollution is the form of pollution which causes people the most concern. It is usually obvious by its effects on the eyes and nostrils and also causes conspicuous toxicity symptoms in vegetation. Nevertheless, some of the most harmful chemicals in the polluted air may not be detectable by smell or any of the other senses. Water pollution is the second most obvious type of pollution, especially when it affects drinking water supplies or causes the death of large numbers of fish. In contrast, soil pollution is often far less conspicuous but it is still very important. As a result of the adsorptive and buffering properties of soil, some pollutants (such as Cu, Pb and PCBs) have long half-lives in the soil and food crops grown on these polluted soils may be affected by some of the pollutants for centuries, even millennia, because soil is difficult and expensive to clean-up (see chapter 2). However, it is a source of comfort that ecosystem and biogeochemical processes are amazingly resilient and that the impact of people on the environment is not as severe as would be expected given the loading of pollutants emitted each year. This resilience is due to the adsorption and detoxification processes operating in soils and sediments, which remove pollutants from circulation and, in many cases, either fix them more or less indefinitely or degrade them to harmless products (see chapter 2). Most of this detoxification of organic chemicals is carried out by microorganisms, which can evolve the ability to synthesize the necessary enzymes for denaturing new chemicals as a result of their rapid rates of reproduction and mutation. However, there is no room for complacency because natural processes for the assimilation and detoxification of pollutants can be overloaded if the rate of pollutant emission and transport is too great.

Given that environmental pollution poses one of the greatest threats to the health and food security of the human race, the need for a greater understanding of it becomes even more important. This book introduces many of the key principles to be considered with regard to the occurrence, behaviour and effects of pollutants in the environments.

References

Allen, J. E. (1992) *Energy Resources for a Changing World*, Cambridge University Press, Cambridge.

Ames, B. N. and Gold, L. S. (1990) *Angem. Chem. Int. Ed. Engl.*, **29**, 1197.

British Medical Association (1991) *Hazardous Waste and Human Health*, Oxford University Press, Oxford.

Carson, R. (1962) *Silent Spring*, Houghton Miflin, Boston.

Holdgate, M. W. (1979) *A Perspective of Environmental Pollution*, Cambridge University Press, Cambridge.

Jones, K. C. (1991) *Environ. Pollut.* **69**, 311.

Meadows, D. H., Meadows, D. L. and Randers, J. (1972) *Limits to Growth*, Universe Books, New York.

Meadows, D. H., Meadows, D. L. and Randers, J. (1992) *Beyond the Limits*, Earthscan Publications, London.

Moldan, B. and Schnoor, J. L. (1992) *Environ. Sci. Technol.* **26**, 14.

Simmonds, I. G. (1987) Energy in human geography: an introduction. *Occasional Publication No. 21*, Department of Geography, University of Durham.

World Resources Institute (1992) *World Resources 1992–93*, Oxford University Press, Oxford.

Air pollution   **Further reading**

Bridgeman, H. (1990) *Global Air Pollution*, Belhaven Press, London.

Elsom, D. (1987) *Atmospheric Pollution*, Blackwell, Oxford.

Environmental chemistry

Manahan, S. E. (1991) *Environmental Chemistry*, Lewis Publishers, Chelsea, Michigan.
Richardson, M. L. (1991) (ed) *Chemistry, Agriculture and the Environment*, Royal Society of Chemistry, Cambridge.

Environmental science

Ellis, D. (1989) *Environments at Risk: case histories of impact assessment*, Springer Verlag, Berlin, Heidelberg.
Masters, G. M. (1991) *Introduction to Environmental Engineering and Science*, Prentice Hall, Englewood Cliffs, New Jersey.
Tolba, M. K., El-Kholy, O. A., El-Hinnawi, E., Holdgate, M. W. and Munn, R. E. (1992) *The World Environment 1972–1992*. UNEP/Chapman and Hall, London.

Pollution

Coughtrey, P. J., Martin, M. H. and Unsworth, M. H. (eds) (1987) *Pollutant Transport and Fate in Ecosystems*, Blackwells, Oxford.
Harrison, R. M. (ed) (1990) *Pollution: Causes, Effects and Control*, Royal Society of Chemistry, London.
Holdgate, M. W. (1979) *A Perspective of Environmental Pollution*, Cambridge University Press, Cambridge.
Murley, L. (ed) (1992) *The National Association of Clean Air Pollution Handbook*, The National Association of Clean Air, Brighton (updated annually).

World resources

World Resources Institute (1992) *World Resources 1992/93*, Oxford University Press, Oxford (earlier editions for 1986, 1987, 1988/89, 1990/91).

Soil science

Ross, S. (1989) *Soil Processes*, Routledge, London.
White, R. E. (1987) *Introduction to the Principles and Practice of Soil Science*, Blackwells, Oxford.

Toxicity

Rodricks, J. V. (1992) *Calculated Risk*, Cambridge University Press, Cambridge.
Harte, J., Holdren, C., Schneider, R. and Shirley, C. (1991) *Toxics A to Z*, University of California Press, Berkeley and Los Angeles.

Wastes

British Medical Association (1991) *Hazardous Waste and Human Health*, Oxford University Press, Oxford.

Water pollution

Clark, R. B. (1989) *Marine Pollution*, Clarendon Press, Oxford.

2 Transport and behaviour of pollutants in the environment

2.1
A basic model of
environmental pollution

The pollution pathway concept put forward by Holdgate (1979) is a very convenient way for studying and appreciating environmental pollution. All pollution events have certain characteristics in common: all involve (i) the pollutant, (ii) the source of the pollutant, (iii) the transport medium (air, water or soil), and (iv) the target (the organisms, ecosystems or items of property affected by the pollutant). These are shown in Figure 2.1.

Varying degrees of sophistication can be added to this simple model including the rate of emission of the pollutant from the source, the rate of transport, chemical and physical transformations which the pollutant undergoes either during transport or after deposition at the target, amounts reaching the target, movement within the target to sensitive

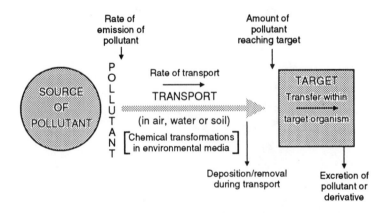

Figure 2.1 A simplified model of environmental pollution (from Holdgate, 1979).

organs, and quantification of the effects on the target. The various components of this model are considered below.

Sources of pollutants can either be discrete point sources or diffuse sources (non-point sources). Even at point sources, fugitive emissions may occur as a result of leakages in an apparently closed system, in addition to overt emission from chimneys or effluent discharge pipes. Some of the major sources and examples of the groups of pollutants likely to be involved are given in Table 2.1.

The following list of the major contaminating uses of land provides an indication of the wide range of possible sources: waste disposal sites, scrapyards, ship and vehicle breaking yards, gas works, petroleum refineries, petrol storage and distribution, petrol service stations, coal mining and coal storage, electricity generation, iron and steel works, metalliferous mining and smelting, metal products fabrication and metal finishing, chemical works, glass-making and ceramics, textile plants, dye works, leather tanneries, timber and timber products treatment works, manufacture of integrated circuits and semi-conductors, food processing, water treatment and sewage works, asbestos works, docks and railway land, paper and printing works, heavy engineering installations, radioactive waste processing, military bases and training areas and the burial of diseased farm livestock (House of Commons Environment Committee, 1990). Although specifically related to land contamination many of these sources also give rise to air and water pollution.

Pollutants have certain intrinsic properties which determine the likely effect that they will have after emission or discharge into the environment. Holdgate (1979) divided these into two types: effect generating properties, such as toxicity in living organisms or corrosion of metals, and pathway determining properties which determine the distance and the rate of dispersion of the pollutant in the environment. These properties, which Holdgate (1979) suggested ought to be evaluated, include: (a) short- and long-term toxicity, (b) persistence, (c) dispersion properties, (d) chemical reactions that the compound undergoes, including its decomposition, (e) tendency to be bioaccumulated in food-chains, and (f) ease of control.

2.3.1   Classification of hazardous substances in the USA

In the USA, hazardous substances have been defined by the Environmental Protection Agency under the authority of the Resource, Recovery and Conservation Act of 1976 and its Hazardous and Solid Wastes Amendments of 1984. Hazardous substances are defined in terms of their

Table 2.1 Major types of sources of environmental pollutants and the environmental media they are transported or reside in

1. *Agricultural sources*
   - Air: Pesticide aerosols, feather dusts, NH_3, H_2S, noxious odours, soil particles
   - Water: Leachates from silage clamps and slurry lagoons, NO^{3-}, HPO_4^{2-}, pesticide spillages, runoff, soil particles, HCs (fuel spillages)
   - Soil: Fertilizers – e.g. As, Cd, Mn, U, V and Zn in some phosphatic fertilizers
             Manures – e.g. As and Cu in pig and poultry manures
             Pesticides – As, Cu, Mn, Pb, Zn, persistent organics (e.g. DDT, Lindane)
             Corrosion of metals – e.g. galvanized metal objects (fencing, troughs, etc.)
             Fuel spillages – HCs
             Burial of dead livestock – pathogenic microorganisms
2. *Electricity generation*
   - Air: CO_x, NO_x, SO_x, UO_x and PAHs from coal, radioisotopes from nuclear fission
   - Water: Heat, biocides from cooling water, soluble B and As compounds and PAHs from ash
   - Soil: Ash, fallout-Si, SO_x, NO_x, heavy metals, coal dust
3. *Derelict gas works sites*
   - Air: VOCs, H_2S, NH_3
   - Water: PAHs, phenols Cu, Cd, As, CN, sulphates
   - Soil: Tars (containing HCs, phenols, benzene, xylene, naphthalene and PAHs), CN, spent Fe oxides, Cd, As, Pb, Cu, sulphates, sulphides
4. *Metalliferous mining and smelting*
   - Air: SO_x, Pb, Cd, As, Hg, Ni, Tl, etc. particulates/aerosols
   - Water: SO_4^{2-}, CN frothing agents, metal ions, tailings (ore minerals e.g. PbS, ZnS, $CuFeS_2$, etc.)
   - Soil: Spoil and tailings heaps – wind erosion, weathering ore particles
             Fluvially dispersed tailings – deposited on soil during flooding, river dredging, etc.
             Transported ore separates – blown from conveyance, etc. onto soil
             Ore processing – cyanides, range of metals
             Smelting – wind-blown dust, aerosols from smelter (range of metals)
5. *Metallurgical industries*
   - Air: Particulates/aerosols: As, Cd, Cr, Cu, Mn, Ni, Pb, Sb, Tl and Zn, VOCs, acid droplets
   - Water: Metal ions, acid wastes and solvents (VOCs) from metal cleaning
   - Soil: Metals in wastes, solvents, acid residues, fallout of aerosols, etc. from casting and other pyrometallurgical processes
6. *Chemical and electronic industries*
   - Air: VOCs, Hg, numerous volatile compounds
   - Water: Waste disposal, wide range of chemicals in effluents, solvents from microelectronics
   - Soil: Particulate fallout from chimneys
             Sites of effluent and storage lagoons, loading, packaging areas
             Scrap and damaged electrical components – PAHs, metals, etc.
7. *General urban/industrial sources*
   - Air: VOCs, particulates, aerosols (e.g. Pb, V, Cu, Zn, Cd, PAHs, PCBs, dioxins, smoke).
             Fossil fuel combustion – CO_x, SO_x, NO_x, As, Pb, U, V, Zn, PAHs
             Bonfires – PAHs, PCDDs, PCDFs, Pb, Cd, etc.
             Cement manufacture – particulates, Ca, SO_4^{2-}, Si, etc.
   - Water: Wide range of effluents, PAHs from soots, Pb, Zn, etc., waste oils – HCs, PAHs, detergents
   - Soil: Pb, Zn, V, Cu, Cd, PCBs, PAHs, dioxins, HCs, dumped cars, etc. – HCs, asbestos, PAHs
8. *Waste disposal*
   - Air: Incineration – fumes, aerosols and particulates (Cd, Hg, Pb, CO_x, NO_x, PCDDs, PCDFs, PAHs)
             Landfills – CH_4, VOCs
             Livestock farming wastes: CH_4, NH_3 H_2S
             Scrapyards – combustion of plastics (PAHs, PCDDs, PCDFs)
   - Water: Landfill leachates, NO_3^-, NH_4^+, Cd, PCBs, microorganisms
             Effluents from water treatment – organic matter, HPO_4^{2-}, NO_3, NH_4^+
   - Soil: Sewage sludge – NH_4^+, PAHs, PCBs, metals (Cd, Cr, Cu, Hg, Mn, Mo, Ni, Pb, V, Zn, etc.)
             Scrapheaps – Cd, Cr, Cu, Ni, Pb, Zn, Mn, V, W, PAHs, PCBs
             Bonfires, coal ash, etc. – Cu, Pb, PAHs, B, As
             Fallout from waste incinerators – Cd, PCDFs, PCBs, PAHs
             Fly tipping of industrial wastes (wide range of substances)
             Landfill leachate – NO_3^-, NH_4^+, Cd, PCBs, microorganisms

Table 2.1 *continued*

---

9. *Transport*

Air: Exhaust gases, aerosols and particulates (e.g. CO_x, NO_x, SO_x, smoke, PAHs, PAN, O_3, PbBrCl, V, Mo)

Water: Spillages of fuels, spillages of transported loads (e.g. hydrocarbons, pesticides, manufactured organic chemicals, wastes in transit – especially marine pollution from oil tanker operations and accidents), road and airport de-icers (e.g. ethylene glycol, various salts), deposition of fuel combustion products, smoke, PAHs, SO_x, NO_x, PbBrCl

Soil: Particulates (PbBrCl, PAHs) acid deposits de-icers, wide range of soluble/insoluble compounds at docks and marshalling yards and sidings, deposition of fuel combustion products, smoke, PAHs, SO_x, NO_x, rubber tyre particles (containing Zn and Cd)

10. *Incidental sources*

Water: Leakage from underground storage tanks e.g. solvents, petrol products

Soil: Preserved wood (e.g. PCP, creosote, etc.), discarded batteries (Hg, Cd, Ni, Zn) fishing and shooting (Pb), galvanized roofs and fences Zn, Cd

All media: Warfare (e.g. fuels, explosives, ammunition, bullets, electrical components, poison gases, combustion products – PAHs), corrosion of metal objects – Cu, Zn, Cd, Pb
Industrial accidents e.g. Bhopal, Séveso, Chernobyl nuclear reactor (wide range of pollutants)

11. *Long range atmospheric transport* (deposition of transported pollutants)

Water and   As, Pb, Cd, Hg, UO_x, Zn, SO_4^{2-}, NO_x, pesticides, PAHs
soil: Wind-blown soil particles with adsorbed pesticides and pollutants

---

PAHs = polycyclic aromatic hydrocarbons; VOCs = volatile organic compounds (see Table 2.2)

ignitability, corrosivity, reactivity and toxicity. In addition, more than 450 wastes are listed as specific substances or types of substances. These are given one of the following letters and a three digit number:

F-type   wastes from non-specific sources
K-type   wastes from specific sources
P-type   acute hazardous wastes
U-type   generally hazardous wastes

2.3.2 *European Community Dangerous Substances Directive*

Although almost any substance present in excess in the wrong place in the environment at the wrong time can cause pollution, chemicals belonging to the following groups of substances are considered to be the major priority concerns. These constitute the EC Dangerous Substances Directive List I (black list) of substances and List II (grey list) (EC, 1976, 1981). Table 2.3 gives the classes of pollutants but, except in the case of 'other' substances, the names of individual compounds are not given.

Table 2.2 USA classification of hazardous substances (from Identification and Listing of Hazardous Wastes Code of Federal Regulations (1986) in Manahan (1984 and 1991))

Waste type	Examples
F-type (waste from non-specific sources)	
F001	Spent halogenated solvents used in degreasing; tetrachlorethane, trichloroethylene and methylene chloride.
F004	Spent non-halogenated solvents: cresols, cresylic acid and nitrobenzene, and the still bottoms from the recovery of these solvents
F007	Spent plating-bath solutions
F010	Quenching bath sludge from oil baths from metal heat treating operations
K-types (hazardous wastes from specific sources)	
K001	Bottoms sediment sludge from the treatment of wastewaters from wood-preserving processes that use creosote and/or pentachlorophenol
K002	Wastewater treatment sludge from the production of chrome yellow and orange pigments
K008	Oven residue from the production of chrome oxide green pigments
K019	Heavy ends from the distillation of ethylene dichloride
K020	Heavy ends (residue) from the distillation of vinyl chloride
K027	Centrifuge residue from toluene diisocyanate production
K043	2,6-Dichlorophenol waste from the production of the herbicide 2,4-D
K047	Pink/red water from TNT operations
K049	Slop oil emulsion solids from the petroleum refining industry
K060	Ammonia lime sludge from coking operations
K067	Electrolytic anode slimes and sludges from primary zinc production
P-type (acute hazardous wastes)	
P003	Acrolein
P013	Barium cyanide
P024	p-Chloroaniline
P050	Endosulfan
P056	Fluorine
P063	Hydrocyanic acid
P065	Mercury fulminate
P081	Nitroglycerine
P095	Phosgene
P105	Sodium azide
P110	Tetraethyl lead
P115	Thallium sulphate
P122	Zinc phosphide
U-type (generally hazardous wastes)	
U001	Acetaldehyde
U021	Benzidine
U032	Calcium chromate
U071	Dichlorobenzene
U115	Ethylene oxide
U133	Hydrazine
U147	Maleic anhydride
U190	Phthalic anhydride
U220	Toluene

Table 2.3 The European Community lists of dangerous substances (EC, 1976, 1981)

List I (129 chemicals specified)
   Chlorinated hydrocarbons
   Chlorophenols
   Chloroanilines and nitrobenzenes
   Polycyclic aromatic hydrocarbons (PAHs)
   Inorganic chemicals (including metals and compounds of As, Cd, Hg, dibutyltin
       compounds and tetrabutyltin)
   Solvents
   Pesticides
   Others (including benzidine, benzyl chloride, benzylidene chloride, chloral hydrate,
       chloroacetic acid, chloroethanol, dibromomethane, dichlorobenzidine, dichloro
       diisopropyl ether, diethylamine, dimethylamine, epichlorohydrin, isopropylbenzene,
       tributyl phosphate, trichlorotrifluoroethane, vinyl chloride, xylenes)

List II
Includes certain substances belonging to the families and groups in List I which do not
   have specified limit values, plus:
   Zn, Cu, Ni, Cr, Pb, Se, As, Sb, Mo, Ti, Sn, Ba, Be, B, U, V, Co, Tl, Te, Ag and their
       compounds
   Biocides and their derivatives not in List I
   Organic silicon compounds
   Inorganic phosphorus compounds
   Non-persistent mineral oils and hydrocarbons of petroleum origin
   Cyanides and fluorides
   NH_3 and NO_3

2.3.3   UK priority list of pollutants

The UK Department of the Environment has established a 'red list' of
priority pollutants and the initial version of this list (1988) is presented in
Table 2.4. It is expected that this list will be added to in due course.

2.3.4   Pesticides

Pesticides are used widely in agriculture and for many other purposes all
over the world with more than 10 000 commercial formulations of around
450 pesticidal compounds currently in use. Their use and dispersion in the
environment has mainly occurred in the last 50 years and they have become
relatively ubiquitous pollutants in human and animal tissues, in soils and
crops, and in groundwater, rivers and lakes, especially in technologically
advanced countries. However, transport in the atmosphere, the oceans and
oceanic food chains has resulted in their wider global distribution. Traces
of several pesticides can be found in the Arctic snows (Welch *et al.*, 1991)
and in Antarctic penguins (Holdgate, 1979).

Pesticides mainly comprise insecticides, herbicides (weed killers) and
fungicides and some of the main types of compounds currently in use are
given in Table 2.5.

Table 2.4   The UK priority red list of pollutants (DOE, 1988)

Mercury and its compounds
Cadmium and its compounds
Gamma-hexachlorocyclohexane
DDT
Pentachlorophenol
Hexachlorobenzene
Hexachlorobutadiene
Aldrin
Dieldrin
Endrin
Polychlorinated biphenyls (PCBs)
Dichloros
1,2-dichloroethane
Trichlorobenzene
Atrazine
Simazine
Tributyltin compounds
Triphenyltin compounds
Trifluralin
Fenitrothion
Azinphos-methyl
Malathion
Endosulfan

Table 2.5   The main types of compounds used as pesticides (from Ross (1989) and Manahan (1991))

Insecticides
    Organochlorines (e.g. DDT, Lindane, Aldrin, and Heptachlor)
    Organophosphates (e.g. Parathion, Malathion)
    Carbamates (e.g. Carbaryl, Carbofuran)

Herbicides
    Phenoxyacetic acids (e.g. 2,4-D; 2,4,5-T, MCPA)
    Toluidines (e.g. Trifluralin)
    Triazines (e.g. Simazine, Atrazine)
    Phenylureas (e.g. Fenuron, Isoproturon)
    Bipyridyls (e.g. Diquat, Paraquat)
    Glycines (e.g. Glyphosate 'Tumbleweed')
    Phenoxypropionates (e.g. 'Mecoprop')
    Translocated carbamates (e.g. Barban, Asulam)
    Hydroxyaryl nitriles (e.g. Ioxynil, Bromoxydynil)

Fungicides
    Non-systemic fungicides
        Inorganic and heavy metal compounds (e.g. Bordeaux Mixture–Cu)
        Dithiocarbamates (e.g. Maneb, Zineb, Mancozeb)
        Pthalimides (e.g. Captan, Captafol, Dichofluanid)
    Systemic fungicides
        Antibiotics (e.g. Cycloheximide, Blasticidin S, Kasugamycin)
        Benzimidazoles (e.g. Carbendazim, Benomyl, Thiabendazole)
        Pyrimidines (e.g. Ethirimol, Triforine)

Table 2.6 Common indoor pollutants and their sources (Adapted from Tolba *et al.* (1992) and Masters (1991))

Pollutant	Source
Formaldehyde	Particle board, plywood, foam insulation, smoking*
NO_2	Gas stoves, kerosene heaters
CO	Kerosene heaters, wood stoves, smoking, vehicles in garages
PAHs	Burning wood, coal or dung, industrial solvents, wood stoves
SO_2	Kerosene heaters
Cl_2	Household bleaches and toilet cleaners
O_3	Photocopiers, laser printers, electrostatic air cleaners
VOCs	Cooking, room deodorizers, cleaning sprays, paints, varnishes, solvents, carpets, furniture
Smoke and other particulates	Smoking, cooking, aerosol sprays, rubber-backed carpets, wood stoves, Pb etc. from erosion of paint emulsions, asbestos and fibrous insulation, decoration, flooring and cement products
Moulds, fungi Viruses	Dampness (humid, cold, and poorly ventilated rooms)
Radon**	Rock, soil, concrete

* Smoking, tobacco/cigarette smoking.
** Radon often naturally occurring.
VOCs, volatile organic compounds; PAHs, polycyclic aromatic hydrocarbons.

2.3.5   Indoor pollution (see also p. 220)

Indoor air pollution in domestic houses is particularly important in considerations of health because people spend a high proportion of their life breathing the air in this environment, especially in their bedrooms. Although many types of pollutant could be present in houses near to sources of industrial pollution, the most common pollutants which are likely to be found in the air of most homes are given in Table 2.6. These substances are dealt with in more detail in various places throughout this book. It must also be remembered that particulate pollutants from the workplace can be transferred to the household environment on clothes. This has been the cause of exposure of workers' families to hazardous materials, such as Pb in the case of smelter workers and asbestos fibres from dockers and workers handling this material, although there are sometimes sources of these pollutants originating in the home (Pb from old paint and asbestos from some old decorating materials). This problem is now recognized in many industrialized countries and health and safety procedures have been introduced to prevent this form of pollutant transfer.

Holdgate (1979) comments that mycotoxins produced in stored food products such as grains and nuts by fungi may possibly have more damaging effects on human health than many anthropogenic pollutants. These fungal moulds are more of a problem in the warm and humid conditions of the tropics than in cooler or dry conditions. Fungal toxins can have both acute and chronic effects, including the onset of cancer; for example the aflatoxin secreted by the fungus *Aspergillus flavus* in peanuts and other stored foods is known to be a very potent carcinogen. However, unlike most environmental pollutants, this pollutant is hardly transported to its target because its source is the fungal mould in the food, its medium is the food, and the act of consuming this food transfers the toxin to its target.

2.4
Physical processes of pollutant transport and dispersion

2.4.1   Transport media

Pollutants emitted from a source are dispersed in the environment in air (e.g. smoke, SO_x, NO_x, and $PbClBr$), in water (e.g. industrial effluents, pesticides, phenols, landfill leachate, NO_3^-) and in soil. Pollutant movement through the soil by pedological processes is often much slower than in other transport media but rapid secondary transport of the pollutant while adsorbed on soil particles can occur when these are carried by the wind or running water. In the context of toxicology, humans and other animals can be exposed to pollutants in the following media: diet (often involving soil–plant pathways), air, water and dust. The transport mechanisms include movement in wind and water, gravity (e.g. movement of particles down the sides of waste heaps into rivers or onto soil), and anthropogenic transport (by land, air or sea) and placement (direct tipping of material).

2.4.2   Transport of pollutants in air

Most air pollutants are discharged into the boundary layer, which is the thin layer of the atmosphere in contact with the earth's surface where the airflow is frequently turbulent due to surface roughness. Many pollutants, especially the larger particulates (1–10 μm) remain in the boundary layer but gases and smaller aerosol particles (< 5 μm) are transferred into the troposphere zone above by vertical movements in thermal plumes, storms and flow over mountains. The troposphere lies immediately above the boundary layer and the air gets cooler with height up to the tropopause which occurs at around 10 km at the poles and 16 km at the equator. Above the tropopause is the stratosphere where there is a reversal in the temperature gradient up to around 50 km. The mesosphere lies above the stratosphere but pollutants do not normally enter this zone (see section 5.1).

The transport of atmospheric pollutants depends on the height they reach in the atmosphere, their particle size and climatic factors. Little transfer of air and pollutants occurs between the northern and southern hemispheres in the troposphere, but some transfer occurs between the troposphere and the stratosphere near the equator and in other places. Explosions, volcanic eruptions and high-flying aircraft inject some pollutants directly into the stratosphere. Gaseous pollutants tend to remain in the stratosphere for a long time due to the lack of washout. However, some pollutant molecules may be degraded by the intense radiation which occurs above the ozone layer (at around 20 km altitude) (Holdgate, 1979).

Air pollutants tend to be transported by the wind and mixed with the surrounding air until their concentration in the turbulent boundary layer is relatively uniform. Dispersion in the horizontal plane is generally unrestricted and occurs more rapidly than vertical mixing. The extent to which air pollutants become diluted after emission is largely controlled by the factors determining the degree of turbulence in the boundary layer and these include: incoming solar radiation, wind speed, cloud cover and land surface roughness (Masters, 1991). The density of air is inversely proportional to its temperature and therefore warm air is less dense and rises, while cooling air becomes denser and descends. However, in the troposphere temperature decreases with increasing height therefore the warm rising air gradually cools and descends again. The temperature of the air also determines how much water vapour it can hold. Descending cool air can become saturated with water vapour and form clouds or fog (Dix, 1981).

The adiabatic lapse rate of the cooling of air masses plays a major part in determining the stability of the air into which pollutants are discharged. Upward movement of air in the troposphere results in expansion and cooling, whereas downward movement leads to compression and warming (Masters, 1991). The adiabatic lapse rate of dry air is 9.76°C/km (or about 1°C/100 m). When the temperature of rising air decreases faster than the adiabatic lapse rate, the air mass becomes unstable and rapid mixing and dilution of pollutants occurs. However, if the temperature decreases more slowly than the adiabatic lapse rate the air will remain stable and the pollutants will concentrate. If there is sufficient moisture in the air for condensation to occur, latent heat will be released and the air mass will cool more slowly. On average, the wet adiabatic lapse rate is about 6°C/km but will be higher at the poles and lower in the tropics (Masters, 1991).

Radiation inversions occur on cold nights (especially in winter) when the temperature of the air above the ground is colder than the ground temperature. They start at ground level around dusk and can move upwards to an altitude of several hundred metres before dispersing the following morning. Since these inversions occur in darkness, no photochemical reactions occur and so primary pollutant gases (such as CO and SO_2) will accumulate and not secondary pollutants (such as O_3, NO_2 and PAN) (see section 6.2).

Subsidence inversions differ markedly from radiation inversions and are associated with high pressure weather systems (anticyclones). At higher elevations they can last for months at a time and are more common in summer than winter. In a high pressure zone, air in the middle descends and moves outwards near the ground, while air at the edges rises. The descending air in the middle of the high pressure system undergoes adiabatic heating to a temperature above that of the air near the ground and results in an inversion of up to several thousand metres' vertical extent that persists for as long as the high pressure system lasts. These inversions are responsible for many regional air pollution problems, such as those of the Los Angeles and San Francisco areas of the south-west USA. A large high pressure zone off the Californian coast remains in position from spring to autumn resulting in clear sunny skies which are conducive to photochemical reactions causing the formation of smogs. There is virtually no washing-out of pollutants due to the lack of rain. The topography in the Los Angeles area further exacerbates the problem because the mountains inland prevent winds from dispersing pollutants away from the area (Masters, 1991).

The amount of air available for the dilution of atmospheric pollutants is dependent on the wind speed and the extent to which the emission can rise into the atmosphere. The mixing depth of the atmosphere is determined by the altitude at which rising adiabatically cooling air would reach the same temperature as the surrounding air. The product of the maximum mixing depth and the average wind speed within this depth is known as the 'ventilation coefficient' (m^2/s) and values of this coefficient of < 6000 m^2/s are regarded as indicative of high potential for pollution (Masters, 1991). Wind speeds increase with height above the ground and the more stable the atmosphere, the higher the wind speed becomes with increasing altitude. The differential between the wind speed at ground level and at higher altitude is exacerbated by the roughness of the land surface (due to surface drag).

Transport of pollutants in chimney plumes (point sources of pollution). Since many atmospheric pollutants are emitted from chimneys (or stacks) (Figure 2.2) it is convenient to use an example of the dispersion of a plume of smoke from a single chimney to illustrate the principal factors involved. As in all other considerations of air pollutant dispersion, the stability of the atmosphere and the speed of the wind are very important, as shown in Figure 2.3.

The nature of the plume from a chimney discharging into a neutrally stable atmosphere (Figure 2.3a) is symmetrical (cone shaped). In contrast, where the atmosphere is unstable there is rapid vertical air movement, both up and down, giving a *looping* plume (Figure 2.3b). A stable atmosphere which restricts the vertical dispersion of a plume will give rise to a *fan-shaped* plume due to limited vertical dispersion with unlimited horizontal dispersion (Figure 2.3c). Where a chimney emission is below an inversion, the resulting *fumigation* will result in increased concentrations of pollutants downwind from the source (Figure 2.3d). When the top of the chimney is above an inversion, vertical mixing is not restricted but downward mixing is limited by the inversion's stable air. This gives rise to *lofting* (Figure 2.3e) which assists in the dilute and disperse approach to

Figure 2.2 Plume of steam from an incinerator with emissions control.

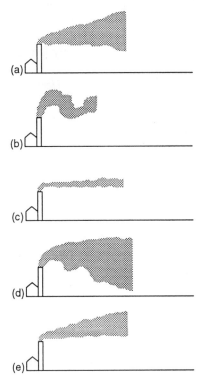

Figure 2.3 Smoke plumes from a chimney under different atmospheric conditions: (a) coning, (b) looping, (c) fanning, (d) fumigation and (e) lofting (adapted from Masters, 1991).

pollution abatement. Over recent decades, high chimneys have been built at major sources of atmospheric pollution, such as coal-burning electricity generating stations and metal ore smelters (e.g. Sudbury in Ontario with a 300 m stack) to discharge the emissions well above inversion layers. Areas immediately downwind from the source are less troubled by atmospheric pollution but long distance tropospheric transport of the pollutants may result in more chronic effects on ecosystems over larger areas, such as acid precipitation and the associated problems of acidification of lakes and die-back of trees (see section 5.2).

It is important to realize that the presence of tall buildings, other chimneys and the urban environment can greatly complicate the pattern of dispersion of atmospheric pollutants from a single chimney. Down-draughts and concentration in urban heat islands can have a marked effect.

Point-source Gaussian model for pollutant transport. Most models of time-averaged concentrations of pollutants down-wind from a point source, such as a chimney, are based on a normal (or Gaussian) distribution curve of the pollutants (see Figure 2.4). The main variables which have to be considered include the rate of emission from the source (assumed to be constant in simple models) and the wind speed. It is also assumed that the pollutant is conservative (i.e. not lost by decay or deposition) and that the terrain is relatively flat. Masters (1991) quotes a model:

$$C(x,y) = \frac{Q}{\pi\mu\sigma_y\sigma_z}\left(\exp\frac{-H^2}{2\sigma_z^2}\right)\left(\exp\frac{-y^2}{2\sigma_y^2}\right)$$

where $C(x,y)$ = concentration at ground level at point x,y (µg/m);

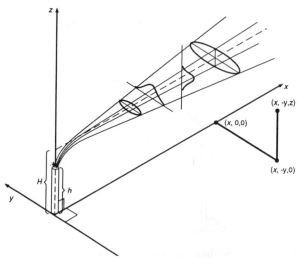

Figure 2.4   A Gaussian model for the dispersion of a plume in the vertical and horizontal directions (reproduced from Turner (1970) in Masters (1991)).

x = distance downwind (m); y = horizontal distance from the plume centreline (m); Q = emission rate of pollutants (μg/s); H = effective stack height (m) ($H=h+\Delta h$, where h=chimney height and Δh=plume rise); μ = average wind speed at the effective height of the stack (m/s); σ_y = horizontal dispersion coefficient (standard deviation) (m); and σ_z = vertical dispersion coefficient (standard deviation) (m).

The concentration of pollutants at ground level is directly proportional to the source strength Q. The ground-level concentrations decrease for higher chimneys (but the relationship is not linear). Although the downwind pollutant concentration appears approximately inversely proportional to wind speed, the relationship is slightly affected by the wind speed which influences the rise of the plume. Higher wind speeds reduce the effective height of the chimney (H) and so the ground-level pollution does not decrease as much as would be expected from the inverse relationship (Masters, 1991). The dispersion coefficients σ_y and σ_z are the standard deviations of the vertical and horizontal Gaussian distributions and are functions of downwind distance and atmospheric stability, and they will increase with distance downwind. Even though this simple model is probably only reliable to $\pm$ 50% in most cases, it is still useful for predicting the fate of pollutant smoke emitted from chimneys.

The rise of the plume after emission is an important factor and is dependent on the buoyancy and momentum of the gaseous emissions and on the stability of the atmosphere. The equation for buoyancy flux (F) is:

$$F = gr^2 v_s(1-T_a/T_s)$$

where F = buoyancy flux (m^4/s^3); g = gravitational acceleration (9.8 m/s^2); r = inside radius of chimney; v_s = chimney gas exit velocity (m/s); T_s = chimney gas temperature (K); and T_a = ambient temperature (K) (Masters, 1991).

In a temperature inversion, the pollutants will reflect off the inversion layer and if the inversion is above chimney height then the basic Gaussian equation must be modified to take account of the restricted vertical dispersion.

Transport of pollutants from non-point sources.

(a) Linear sources. Dispersion from a linear source, such as a major road or a burning of stubble in a farm field, can be described by a simple equation assuming an infinitely long source and that the wind blows perpendicular to the line. The concentration at a point x metres away from the line of emission is given by:

$$C(x) = \frac{2q}{\sqrt{2\pi}\,\sigma_z u}$$

where u = wind velocity (m/s) and q = emission rate per unit of distance along an infinite line source (g/m/s) (Masters, 1991).

(b) Area sources.   The simplest way to deal with an area source, such as a city, is to use a box to represent the airshed over the area:

$$\left(\begin{array}{c} \text{rate of change of} \\ \text{pollution in box} \end{array} \right) = \left(\begin{array}{c} \text{rate of pollution} \\ \text{entering box} \end{array} \right) + \left(\begin{array}{c} \text{rate of pollution} \\ \text{leaving box} \end{array} \right)$$

$$LWH \frac{dC}{dt} = q_s LW + WHuC_{in} - WHuC$$

where C = concentration of pollutant in the airshed; C_{in} = concentration in incoming air; L = length of airshed; q_s = emission rate per unit area W = width of airshed; H = mixing height and u = average windspeed against side of box (Masters, 1991).

(c) Indoor air pollution.   Models of indoor air pollution can be modified to take account of the decay of the pollutant:

$$\left(\begin{array}{c} \text{Rate of} \\ \text{increase in} \\ \text{the box} \end{array} \right) = \left(\begin{array}{c} \text{Rate of} \\ \text{pollution} \\ \text{entering the} \\ \text{box} \end{array} \right) - \left(\begin{array}{c} \text{Rate of} \\ \text{pollution} \\ \text{leaving the} \\ \text{box} \end{array} \right) - \left(\begin{array}{c} \text{Rate of} \\ \text{decay in the} \\ \text{box} \end{array} \right)$$

$$V \frac{dC}{dt} = S + C_a IV - CIV - KCV$$

where V = volume of conditioned space in building (m^3/air change); I = air exchange rate (air changes/h); S = pollutant source strength (mg/h); C = indoor concentration (mg/m^3); C_a = ambient concentration (mg/m^3); and K = pollutant decay rate or reactivity (l/h).

A steady state solution is found by setting $dC/dt = 0$

$$C = \frac{S/V + C_a I}{I + K}$$

A general solution is:

$$C(t) = \frac{S/V + C_a I}{I + K} (I - e) + C(0) e$$

where $C(0)$ = the initial concentration in the building.

In the case of conservative pollutants, such as CO and NO where $K=0$, if the ambient concentration is negligible and the initial concentration is 0, the equation will be:

$$C(t) = \frac{S}{IV}(1 - e^{-It})$$

With the exception of CO_2, few other pollutants are accumulating in air; most are transported and then deposited or removed from the air by various mechanisms (Masters, 1991).

2.4.3 Some important types of reactions which pollutants undergo in the atmosphere

(a) *Oxidation*
Examples:
 (i) Oxidation of CO to CO_2 (see section 6.2)
      This occurs as a result of CO reacting with the hydroxyl radical HO·

$$CO + HO· \rightarrow CO_2 + H·$$

   This produces the hydroperoxyl radical

$$O_2 + H· + M \rightarrow HOO· + M$$

  HO· is regenerated by

$$HOO· + NO \rightarrow + NO_2$$

$HOO· + HOO· \overset{-O_2}{\rightarrow} H_2O_2$ which dissociates photochemically to HO·

(ii) Oxidation of SO_2 to SO_3^{2-} and H_2SO_4 (see section 5.2)

   Either $HO· + SO_2 \rightarrow HOSO_2· \rightarrow H_2SO_4$

   or   ½ O_2 (aq) + SO_2(aq) + $H_2O \rightarrow H_2SO_4$

(b) *Photochemical reactions*
 (i) Photodissociation (see section 5.1)

$$NO_2 + hv \rightarrow NO + O$$

(ii) Formation of PAN (peroxyacetyl nitrate) (see section 5.1)

2.5
Transport of pollutants in water

Both the atmosphere and water are fluid media and have many properties in common. The transport and dispersion of pollutants in the aquatic environment is controlled by advection (mass movement) and mixing or diffusion (without net movement of water) (Hewitt and Harrison, 1986). Also, in common with the situation in the atmosphere, the vertical movement of water is often restricted. In large water bodies this is due to stratification, caused by differences in temperature and density or salinity, but in rivers this is due to limited depth. However, unlike the atmosphere, pollutants tend to accumulate in lakes and seas and in the sediments at the bottom of these bodies of water.

As a result of the similarities in the factors governing the dispersion of pollutants of both air and waters, models for both media use a Gaussian plume equation. For rivers, the simplest model for changes in concentration downstream from a point source assumes an exponential decay in concentration:

$$C_x = C_0 e^{-kt}$$

where C_x is the concentration at point x, C_0 is the concentration at the point of discharge, k is the decay rate and t is the time taken for flow from the point of discharge to point x. Other approaches to modelling river pollution involve dividing the river into a series of separate reaches which are assumed to behave as stirred tank reactors. The major dispersive mechanism is the 'aggregated dead zone' (ADZ) phenomena in the river between the two sampling points and the model is a combination of the stirred tank reactor and a factor to account for the advection component of dispersion (Hewitt and Harrison, 1986). ADZ modelling is discussed in detail by Young (1990).

Aquifers are water-bearing strata beneath the land surface and contain groundwater. If soluble pollutants reach the groundwater in an aquifer they may be transported considerable distances and pose a risk to human health since water supply wells remove water from aquifers and groundwater eventually enters surface waters. Although water occurs at depth in soils (the unsaturated groundwater zone) it is the saturated zone which is of most importance. The amount of water stored in an aquifer is dependent on the volume of material and its porosity (percentage of its volume occupied by voids filled with groundwater). Clays have a high porosity (45%) but the diameter of the voids is small so that water is held by capillarity and does not flow under gravity. In contrast, gravel has a lower porosity (25%) but the voids are larger and so the groundwater moves easily under the influence of gravity or a hydrostatic head. For example, typical rates of flow in gravel are 10^{-2} to 10^{0} cm/s and 10^{-5} to 10^{-9} cm/s in silty clays and clays (which are virtually impermeable).

Flow through an aquifer is described by Darcy's Law:

$$Q = KA\frac{dh}{dL}$$

where Q is the volume flux, K is the hydraulic conductivity, a constant determined for each material, A is the cross-sectional area of the aquifer and dH/dL is the hydrostatic head (or hydraulic gradient) which is the driving force.

When the groundwater in an aquifer becomes polluted it is important to be able to determine the movement of the pollutant within the aquifer. As in the case of atmospheric pollution, the pollutant is considered to form a plume and therefore it is the shape and size of the plume which are of particular importance. The parameters involved are: the velocity of the groundwater, the permeability of the aquifer material, the adsorptive properties of the aquifer and the chemical properties of the pollutant.

It is important to recognize that some of the most serious pollution problems in fresh waters involve substances which are not considered to pose a high toxicity risk. Firstly, organic matter which can be easily oxidized biochemically by microorganisms is a major water pollution problem because it can lead to the removal of most of the dissolved oxygen in the water by microorganisms leaving insufficient for fish and other aquatic animals (as discussed in section 2.5.1). The second is nitrate (NO_3^-) pollution. Nitrates occur naturally in soils and waters as a product of the microbial mineralization of dead plant and animal tissues in the soil and forms a very important stage in the nitrogen cycle because NO_3^- is the main form of nitrogen taken up by plants. Ammonium ions from the mineralization process, or from fertilizers such as NH_4NO_3, tend to be rapidly nitrified to NO_3^-. There has been a marked increase in the NO_3^- content of both surface waters and groundwaters over the last 50 years and this is due to the increased use of nitrogen fertilizers. Owing to the beneficial conditions under the EC Common Agricultural Policy (CAP), it has been economic for farmers to apply up to 300 kg N/ha or more to high yielding crops. Since NO_3^- anions are not adsorbed in soils they are leached down the soil profile, so a considerable proportion of the nitrogen applied in fertilizers is leached into groundwaters or surface waters. The effects of excess NO_3^- on human health are not fully understood, but the possibility of them contributing to the synthesis of carcinogenic nitrosamines in the human body has been considered. The maximum allowable NO_3^- concentration in potable (drinking) waters is 50 mg NO_3/l in the UK, but some water supplies in intensive arable regions exceed this value. Peak winter values can frequently exceed 100 mg NO_3/l in rivers draining intensively farmed arable areas, such as much of East Anglia in the UK, and some mathematical models have predicted levels of < 150 mg NO_3/l (Crathorne and Dobbs, 1990). However, recent policies to protect aquifer recharge areas, campaigns to encourage farmers to use nitrogen fertilizers more efficiently, and falling prices for crop products have helped to slow down the NO_3^- increase in waters in many parts of the EC.

2.5.1 *Biochemical processes in waters (involving microorganisms)*

Oxidation of organic carbon compounds

$$\{CH_2O\} + O_2 \rightarrow CO_2 \text{ and } H_2O$$

Oxidation of ammonia

$$NH_4^+ + 2O_2 \rightarrow 2H^+ + NO_3^- + H_2O$$

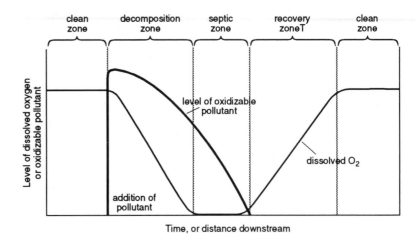

Figure 2.5 Curve of the decrease in oxygen concentration resulting from the addition of oxidizable pollutants to a stream. (From Manahan, S.E., *Environmental Chemistry*, 5/e, Lewis Publishers, Boca Raton, Florida, 1991. With permission.)

The first of these two reactions requires the presence of dissolved oxygen and temperatures above 0°C for oxidation of carbon compounds and > 4°C for oxidation of NH_3 (Fish, 1992). This requirement for oxygen is called the *biochemical oxygen demand* (or biological oxygen demand), usually abbreviated to BOD. If there is an excess of oxidizable organic matter in a river or pond, arising from a discharge of an effluent such as liquid manure slurry from a farm, the bacteria carrying out the oxidation may utilize all the available dissolved oxygen causing an acute shortage of oxygen for fish, which then die from asphyxiation. The BOD of a stream or river is measured by determining the quantity of oxygen utilized by aquatic microorganisms over a 5-day period. Typical values of BOD are < 3 mg/l for Class 1A rivers in the UK (the least polluted class), < 5 mg/l for Class 1B, 9 mg/l for Class 2 (more polluted and only suitable for potable supply after advanced treatment) and 17 mg/l for Class 3 (poor quality water with few fish present) (Fish, 1992). The effect of an input of oxidizable organic pollutant into a stream is shown in Figure 2.5.

Other chemical reactions involve:

1. Reduction of nitrate by microorganisms ($\rightarrow N_2$ and some N_2O) in absence of dissolved O_2.
2. Hydrolysis of pollutants, such as pesticidal esters, amides and organophosphate esters, by microorganisms.
3. Dehalogenation of certain chlorinated compounds.
4. Hydroxylation of aromatic ring compounds.
5. Precipitation of metals (determined by solubility products).
6. Complexation of metals (chelating ligands such as NTA in polluted water may mobilize metals adsorbed on sediments).

2.6.1   *The composition and physico-chemical properties of soils*

2.6
**The behaviour of
pollutants in the soil**

The concentrations of pollutants in moving air or rivers tend to be diluted fairly rapidly due to mixing and dilution but in the case of soil many pollutants tend to accumulate. Soils act as a sink for pollutants due to adsorption processes which bind inorganic and organic pollutants with varying strengths to the surfaces of soil colloids. This adsorption inhibits the leaching of pollutants down the soil profile to the water table, reduces their bioavailability to plants and, in the case of organic pollutants, affects their rate of decomposition. In many cases, the reactions occurring in soils are similar to those in sediments but usually only the surface layer of sediments has oxic conditions. The major thickness of sediments has anoxic conditions which resemble conditions in waterlogged (gleyed) soils.

Soils comprise a mixture of organic, mineral, gaseous and liquid constituents inhabited by a wide range of microorganisms which catalyse many important reactions. Organic matter in soils includes decomposing plant material and humic compounds which have been synthesized by the action of microorganisms on residues of plant material. The mineral constituents of soils can include: particles of weathering rock and discrete rock forming minerals (such as quartz), clay minerals, hydrous oxides of Fe, Al and Mn, and calcite. The humic substances, clay minerals and hydrous oxides are bonded together in various ways and jointly form the colloidal adsorption complex which plays a very important role in determining the behaviour of pollutants.

The liquid and gaseous phases in soils occupy the system of pores created by the voids between the aggregated solid particles. The aqueous soil solution containing ions and soluble organic compounds forms the liquid phase. The gaseous phase is similar in composition to the atmosphere above the soil surface except that the concentration of CO_2 is often more than eight times higher in the soil air. Some organic compounds are lost from the soil by volatilization and these vapour phases will also be present in the soil air. The relative proportions of water and gases in the soil pores have a very important effect on soil physico-chemical properties (redox and pH), the soil biomass and plant growth. Under conditions of prolonged waterlogging the supply of O_2 is rapidly exhausted. Anaerobic conditions rapidly develop and anaerobic microorganisms replace aerobic organisms and catalyse the chemical reduction of compounds, such as the hydrous oxides of Fe^{3+}, Al^{3+} and Mn^{4+} together with SO_4^{2-} and organic compounds. The decomposition of organic pollutants is strongly affected by the redox conditions and the predominant types of microorganisms.

The concentration of ions in the soil solution is determined by the interacting processes of oxidation, reduction, adsorption, precipitation and desorption. When pollutants reach the soil surface they are either adsorbed, with varying strengths on the colloids at the surface of the topsoil, or are washed down through the surface layer into the soil profile in rainwater or snow melt. Soluble pollutants will infiltrate into the topsoil in the system of pores where the adsorption of ions occurs. Insoluble compounds will accumulate on the surface and hydrophobic organic molecules will bind to sites on soil organic matter at the soil surface. These substances become incorporated into the topsoil and deeper profile during mechanical soil movement or down dessication cracks while being adsorbed on soil particles. Some organic pollutant molecules on the soil surface will undergo photolytic decomposition due to exposure to UV wavelengths in daylight.

Several different types of adsorption reaction can occur between the surfaces of organic and mineral colloids and the pollutants. The extent to which the reactions occur is determined by the composition of the soil (especially the amounts and types of clay minerals, hydrous oxides and organic matter), the soil pH, redox status, and the nature of the contaminants. The more strongly pollutants are adsorbed, the less likely they are to be leached down the soil profile or to be available for uptake by plants. Ionic pollutants such as metals, inorganic anions and certain organic molecules, such as the bipyridyl herbicides (e.g. Paraquat), are adsorbed onto soil colloids. Non-ionic organic molecules, which include hydrocarbons, most organic micropollutants and pesticides, are adsorbed onto humic polymers by both chemical and physical adsorption mechanisms. However, some organic pollutants, such as solvents, tend to be relatively easily leached in regions where there is a marked excess of precipitation relative to evapotranspiration. In many cases, adsorption is a necessary preliminary stage in the decomposition of organic pollutant molecules by bacterial extracellular enzymes.

Soil organic matter plays an important role in cation exchange reactions and in the formation of complexes (mostly chelates) with trace metals. Low molecular weight, soluble organic molecules can form stable complexes with metals which are mobile in the soil solution and bioavailable due to the protection of the metal from adsorption on soil colloid surfaces by the organic ligand. However, the more highly polymerized, solid state humus acts as a major adsorbent for metals through the formation of chelates and thus renders them immobile and much less bioavailable.

In addition to the adsorption/desorption processes occurring in soils, the wide range of microorganism species present also have important effects on the behaviour of pollutants. Microorganisms, such as *Thiobacillus* spp. catalyse the oxidation of sulphides. In the case of pollution by tailings from metalliferous mining, particles of ore minerals in the soil, such as PbS, ZnS, and $CuFeS_2$, become oxidized, releasing metal cations such as Pb^{2+}, Cu^{2+}, Zn^{2+} and Cd^{2+} into the soil solution where they undergo adsorption reactions. Sulphide oxidation also causes an increase in soil acidity unless there is a sufficient concentration of carbonates present to buffer it. The decrease in soil pH will diminish the extent of metal adsorption and cause an increase in the concentrations of metal in the soil solution which can be leached down the soil profile or taken up by plant roots. Bacteria do not tolerate highly acid conditions so plant debris and organic pollutant decomposition will be inhibited at low pH. However, many fungal species are able to tolerate strongly acid conditions although the end products of organic matter decomposition may differ from those of the bacteria involved in humification.

Some microorganism species can methylate elements such as As, Se and Hg into volatile forms (e.g. CH_3Hg^+) which then diffuse into the atmosphere as part of the gaseous exchanges of soil and atmospheric gases.

2.6.2   Cation and anion adsorption in soils

Ion exchange refers to the exchange between the counter-ions balancing the surface charge on the soil colloids and the ions in the soil solution. Negative charges on soil colloids are responsible for cation exchange. The extent to which soil constituents can act as cation exchangers is expressed as the cation exchange capacity (CEC), measured in $cmols_c/kg$ (previously in milli. equivs/100 g). Some examples of the typical CEC values for soil colloidal constituents are (Ross, 1989):

	$cmols_c/kg$
Soil organic matter	150–300
Kaolinite (clay)	2–5
Illite (clay)	15–40
Montmorillonite (clay)	80–10
Vermiculite (clay)	150
Hydrous oxides (Fe, Al, Mn)	4

Soil organic matter has a higher CEC than other soil colloids and plays a very important part in adsorption reactions in most soils even though it is normally present in much smaller amounts (1–10%) than clays (<80%). Sandy soils with low contents of both organic matter and clay tend to have low adsorptive capacities and pose a threat for contaminants infiltrating down to the water table.

The negative charges on the surfaces of soil colloids are of two types: (a) permanent charges resulting from the isomorphous substitution of a clay mineral constituent by an ion with a lower valency, and (b) the pH-dependent charges on oxides of Fe, Al, Mn, Si and organic colloids which are positive at pHs below their isoelectric points and negative above their isoelectric points. Hydrous Fe and Al oxides have relatively high isoelectric points (> pH 8) and so tend to be positively charged under most conditions whereas clay and organic colloids are predominantly negatively charged under alkaline conditions. With most colloids, increasing the soil pH, at least up to neutrality, tends to increase their CEC. Humic polymers in the soil organic matter fraction become negatively charged due to the dissociation of protons from carboxyl and phenolic groups.

The concept of cation exchange implies that ions will be exchanged between the soil solution and the zone affected by the charged colloid surfaces (double diffuse layer). The relative replacing power of any ion on the cation exchange complex will depend on its valency, its diameter in hydrated form and the type and concentration of other ions present in the soil solution. With the exception of H^+, which behaves like a trivalent ion, the higher the valency, the greater the degree of adsorption. Ions with a large hydrated radius have a lower replacing power than ions with smaller radii. For example, K^+ and Na^+ have the same valency but K^+ will replace Na^+, owing to the greater hydrated size of the Na^+ ion. The commonly quoted relative order of replaceability on the cation exchange complex of metal cations is:

$$Li^+ = Na^+ > K^+ = NH_4^+ > Rb^+ > Cs^+ > Mg^{2+} > Ca^{2+} > Sr^{2+}$$
$$= Ba^{2+} > La^{3+} = H^+(Al^{3+}) > Th^{4+}$$

For individual soil constituents, the order of replacement of trace metals is (Alloway, 1990):

Montmorillonite clay: Ca > Pb > Cu > Mg > Cd > Zn
Ferrihydrite: Pb > Cu > Zn > Ni > Cd > Co > Sr > Mg
Peat: Pb > Cu > Cd = Zn > Ca

Anion adsorption occurs when anions are attracted to positive charges on soil colloids. As stated above, hydrous oxides of Fe and Al are usually positively charged and so tend to be the main sites for anion exchange in soils. In general, most soils tend to have far smaller anion exchange capacities than cation exchange capacities. Some anions, such as NO_3^- and Cl^-, are not adsorbed to any marked extent but others, such as HPO_4^{2-} and $H_2PO_4^-$, tend to be strongly adsorbed. Some organic pesticides, such as the phenoxyalkanoic acid herbicides, exist as anions at normal soil pHs and are adsorbed to a limited extent by hydrous oxides and by H-bonding to humic polymers.

Specific adsorption is a stronger form of adsorption, involving several heavy metal cations and most anions which form partly covalent bonds with surface ligands on adsorbents, especially hydrous oxides of Fe, Mn and Al. This adsorption is strongly pH specific and the metals and anions which are most able to form hydroxy complexes are adsorbed to the greatest extent. The order for the decreasing strength of specific adsorption of selected heavy metals is:

Cd > Ni > Co > Zn >> Cu > Pb > Hg

Coprecipitation of metals with secondary minerals, including the hydrous oxides of Fe, Al and Mn, is an important adsorptive mechanism in soils with fluctuating moisture status. Cu, Mn, Mo, Ni, V and Zn are coprecipitated in Fe oxides, and Co, Fe, Ni, Pb and Zn are coprecipitated in Mn oxides. Precipitation of Fe(III) is initially in the form of gelatinous ferrihydrite [$Fe_5(O_4H_3)_3$] which gradually dehydrates with ageing to more stable forms, such as goethite. Ferrihydrite is more likely to be subsequently dissolved again through a decrease in Eh or pH than goethite. Ferrihydrite coprecipitates other ions and, as a result of its large surface area, acts as a scavenger sorbing both cations, such as heavy metals, and anions, especially HPO_4^{2+} or $H_2PO_4^+$ and AsO_4^{3-}. Pyrite (FeS_2) forms in severely reducing conditions when sulphate becomes reduced to sulphide, producing H_2S which then reacts with Fe^{2+} to form FeS and FeS_2. The oxidation of sulphides, such as pyrite, causes marked acidification of soils. Specialized bacteria, such as *Thiobacillus ferrooxidans* and *Metallogenum* spp. are involved in the transformations of iron and manganese, respectively. Iron and Mn oxides occur as coatings on soil particles, fillings in voids and as concentric nodules. The oxide coatings are usually intimately mixed with the clay and humus colloids and, although mineralogically distinct, form part of the clay-sized fraction. The trace metals normally found coprecipitated with secondary minerals in soils are (Sposito, 1983):

Fe oxides   V, Mn, Ni, Cu, Zn, Mo
Mn oxides   Fe, Co, Ni, Zn, Pb
Ca carbonates   V, Mn, Fe, Co, Cd
Clay minerals   V, Ni, Co, Cr, Zn, Cu, Pb, Ti, Mn, Fe

When reducing conditions cause the dissolution of hydrous Mn and Fe oxides, the concentrations of several other elements in the soil solution are likely to increase. Cobalt, Ni, Fe, V, Cu and Mn are generally more bioavailable from gleyed (periodically waterlogged) soils than from freely drained soils on the same parent material. However, B, Co, Cu, Mo and Zn do not undergo redox reactions themselves but are coprecipitated by the hydrous oxides.

Coprecipitation of trace metals on carbonates (mainly $CaCO_3$) is very important on semi-arid soils and in soils formed on limestone. In the case of Cd the precipitation of $CdCO_3$ can be accompanied by the chemisorption of Cd, where it replaces Ca in the calcite crystal.

2.6.3 Adsorption and decomposition of organic pollutants

Non-ionic and non-polar organic pollutants are normally adsorbed on soil humic material. Since most soil organic matter is found in the surface horizon, there is a tendency for these pollutants to be concentrated in the topsoil. Migration of organic contaminants down the profile only occurs to any marked extent in highly permeable sandy or gravelly soils with low organic matter contents.

Most organic pollutants are relatively insoluble and do not move down the soil profile but chlorinated solvents tend to be leached fairly rapidly down the profile of most soil types, including peats. Some pesticides also tend to be relatively easily leached, such as the herbicide atrazine (p. 220). In general, apart from solvents and certain pesticides, most organic pollutants that reach water courses after being in contact with soils have been transported to the water course while adsorbed on eroded soil particles and not leached through the soil profile.

Adsorption of organic pollutants depends on their surface charge and aqueous solubility, both of which are affected by the soil pH. Adsorption of non-polar organic contaminants onto soil organic matter will not occur in the presence of oils. Microbially synthesized surfactants can help to accelerate the rate of degradation of hydrocarbon oils in contaminated soils. For many organic pollutants, adsorption onto soil colloids and the presence of water are important catalysts of organic micropollutant degradation, which can be of two types (Ross, 1989):

(a) non-biological degradation: includes hydrolysis, oxidation/reduction volatilization and photodecomposition
(b) microbial decomposition: many pesticide decomposition processes have some biological contribution – there is an initial time lag, while the microorganisms become adapted to the pesticide substrate

In most cases, non-biological degradation processes, such as photo-decomposition and volatilization, occur at the same time as microbially catalysed reactions. The range of factors affecting the degradation of organic contaminants by microorganisms include: soil pH, temperature, supply of oxygen and nutrients, the structure of the contaminant molecules, their toxicity and that of their intermediate decomposition products, the water solubility of the contaminant and its adsorption to the soil matrix (and therefore the organic matter content of the soil). Adsorption tends to decrease and volatilization increase with increasing temperature.

Organic pollutants are decomposed by soil microorganisms but the rate at which this occurs will depend on the nature of the pollutant (its toxicity to microorganisms and that of its decomposition products), the genotype of the microorganisms (whether they have become adapted to decomposing the particular pollutant), the pH and nutrient status of the soil, and its adsorptive properties. Many organic pollutants are more rapidly decomposed after they have been adsorbed onto the soil organic matter. Decomposition of pollutants by microorganisms is brought about by extracellular enzymes and involves an initial lag period while the microorganisms become adapted to the new substrate. Some xenobiotic organochlorine molecules, such as DDT, PCBs and PCDDs, are generally regarded as being highly persistent in soils with residence times of at least

10 years. They have a very slow decomposition rate because the carbon–chlorine bond is not found in nature and so most microorganism species do not possess the enzymes to break this bond. Nevertheless, some species of bacteria and fungi have rapidly evolved the ability to decompose chlorinated organic compounds and these are being utilized in the bioremediation of contaminated land.

Chlorinated solvents are relatively mobile in soils and some tend to be degraded quite rapidly. Studies of the leaching and degradation behaviour of chlorinated solvents in three Dutch soils showed that 1,4-dichlorobenzene was completely degraded under aerobic conditions and did not reach the groundwater. Chloroform, and to a lesser extent 1,1,1-trichloroethylene, were only poorly degraded under both aerobic and anaerobic conditions but trichloroethylene and tetrachloroethylene were more completely degraded. With the exception of 1,4-dichlorobenzene, all the other compounds broke through into the groundwater at 1 metre depth (Loch *et al.*, 1986). Other workers have found that both chlorinated aliphatics and aromatics are leached into groundwaters. Volatilization appears to be the major route for the removal of chlorinated aliphatics from soils, while degradation is not very significant. However, chlorinated benzenes tended to be degraded rather than volatilized (Loch *et al.*, 1986).

Non-biological decomposition of organic pollutants includes photodecomposition when exposed to the UV spectrum of daylight. This usually requires the pollutant to be adsorbed on the surface of soil colloids and obviously only affects the pollutants present on the surface of the topsoil exposed to sunlight. Other non-biological decomposition reactions include oxidation, reduction, hydrolysis and methylation.

Oxidation of organic pollutants occurs by the action of oxygenase enzymes secreted by microorganisms. In alkane hydrocarbons the initial step in this oxidation is the conversion of a terminal CH_3 group to a CO_2H group (Figure 3.1). Aromatic rings are cleaved by the addition of OH to adjacent carbon atoms. The bacterium *Cunninghamella elegans* is known to attack aromatic ring molecules, such as naphthalene. It should be emphasized that the decomposition products of some organic molecules are more toxic to soil microorganisms, animals and humans than the initial compound. For example, the microbial oxidation products of the PAH molecule – benzo-[a]-pyrenes – are carcinogenic because they bind to cellular DNA (see section 6.2.5).

As stated earlier, many pollutants are sorbed by soil colloids and so in many cases little transport occurs within the soil profile, except down desiccation cracks or worm channels while adsorbed to soil particles. Soil particles can also be transported long distances (thousand of kilometres) and a brown snow event in Arctic Canada caused by soil particles transported from China was found to have brought about a significant degree of pollution by pesticides (Welch *et al.*, 1991). Soil particles washed into water courses become part of the sediment load of the stream and may eventually be deposited on the stream or river bed, or in lakes and estuaries where they may undergo reducing reactions which lead to the solubilization of ions and molecules sorbed by hydrous oxides.

However, several solvents such as chloroform, 1,1,1-trichloroethylene and tetrachloroethylene can leach through soil profiles and reach aquifers (Loch *et al.*, 1986). The soil's hydraulic conductivity will be an important factor in determining the rate at which this will occur.

A substantial number of pesticide compounds and their derivatives are present in most agricultural soils because different compounds are used for different pests and crops and new products are continually being introduced due to the development of pest resistance to old formulations. There is therefore the possibility of interactions between pesticides and differing behaviour in the soil and the groundwater. The EC maximum acceptable concentration of pesticides in drinking water is 0.1 μg/l but this is often exceeded (however it is seldom more than 1 μg/l). The pesticides most frequently causing problems with drinking water quality are the herbicides: atrazine, simazine, mecoprop and isoproturon, and carbamate and chloro-propionate insecticides. Water pollution by pesticides is exacerbated by preferential flow down through fissures (desiccation cracks and subsoiling fissures). The concentration of soil-acting pesticide in the soil solution will be thousands of times greater than the maximum permissible concentration in drinking water (Foster *et al.*, 1991). Typical rates of application of pesticides in the UK are around 0.2–5 kg/ha on agricultural land but much higher rates are used for non-agricultural applications of herbicides, e.g. defoliation in forestry firebreaks, and weed control on railways, airfields and highways.

2.7
Concluding remarks

This chapter has attempted to introduce the general principles relating to the behaviour of pollutants in the environment. More detailed examples relating to specific pollutants are given in chapters 5, 6 and 7. However, since this book is focused on the pollutants themselves less coverage is given to their transport.

References

Alloway, B. J. (1990) Chapter 2 in Alloway, B. J. (ed) *Heavy Metals in Soils*, Blackie, Glasgow.

Alloway, B. J. (1992) Chapter 5, in Harrison, R. M. *Understanding Our Environment* (2nd edition), Royal Society of Chemistry, Cambridge.

Crathorne, B. and Dobbs, A. J. (1990) Chapter 1 in Harrison, R. M. (ed) *Pollution: Causes, Effects and Control* (2nd edition), Royal Society of Chemistry, Cambridge.

Department of the Environment (DOE) (1988) Inputs of dangerous substances to water: proposals for a unified system of control. The Government's consultative proposals for a unified system of tighter controls over the most dangerous substances entering aquatic environments. *The Red List*, July.

Dix, H. M. (1981) *Environmental Pollution*, John Wiley and Sons, Chichester.

Elsom, D. (1987) *Atmospheric Pollution*, Blackwell, Oxford.

European Economic Community (1976) *Directive on pollution caused by certain dangerous substances discharged into the aquatic environment of the community,* 76/464/EEC, EEC, Brussels.

European Economic Community (1981) *Directive on pollution caused by certain dangerous substances discharged into the aquatic environment of the community,* 76/464/EEC, Pollution reduction programmes for List II substances, EEC, Brussels.

Fish, H. (1992) Chapter 3 in Harrison, R. M. (ed) *Understanding Our Environment* (2nd edition), Royal Society of Chemistry, Cambridge.

Foster, S. S., Chilton, P. J. and Stuart, M. E. (1991) *J. Inst. Water Env. Man,* 186.

Hewitt, C. N. and Harrison, R. M. (1986) Chap 1 in Hester, R. E. (ed) *Understanding Our Environment* (1st edition), Royal Society of Chemistry, London.

Holdgate, M. W. (1979) *A Perspective of Environmental Pollution,* Cambridge University Press, Cambridge.

House of Commons Environment Committee (1990) 1st Report, *Contaminated Land,* HMSO, London.

Identification and Listing of Hazardous Wastes (1986) Code of Federal Regulations **40**, (July 1) Part 261, US Government Printing Office, Washington DC, pp. 359–408.

Loch, J. P. G., Kool, H. J., Lagas, P. and Verheul, J. H. A. M. (1986) in Assink, J. W. and van den Brink, W. J. (eds) *Contaminated Soil,* Martinus Nijhoff Publishers, Dordrecht, p. 63.

Manahan, S. E. (1984) *Environmental Chemistry* (4th edition), Brooks/Cole Publishing Company, Monterey, California.

Manahan, S. E. (1991) *Environmental Chemistry* (5th edition), Lewis Publishers, Chelsea, Mich.

Masters, G. M. (1991) *Introduction to Environmental Engineering and Science,* Prentice Hall, Englewood Cliffs, NJ.

Ross, S. (1989) *Soil Processes,* Routledge, London.

Sposito, G. (1983) in Thornton, I. (ed) *Applied Environmental Geochemistry,* Academic Press, London.

Tolba, M. K., El-Kholy Osama, A., El-Hinnawi, E., Holdgate, M. W., McMichael, D. F. and Munn, R. E. (1992) (eds) *The World Environment 1972–1992.* UNEP, Chapman and Hall, London.

Welch, H. E., Muir, D. C. G. and Lemoine, B. M. (1991) *Environ. Sci. Technol.* 280.

Young, P. C. (1990) Quantitative Systems Methods in Evaluation of Environmental Pollution Problems in Harrison, R. M. (ed) *Pollution: Causes, Effects and Control,* (2nd edition) Royal Society of Chemistry, Cambridge.

3 Toxicity and risk assessment of environmental pollutants

3.1
Basic principles of toxicology

Human and other animals are exposed to chemicals via water, air, soils, dusts and their diets. These chemicals enter the body by ingestion (mainly in the diet and in water, but also on hands, in soil on vegetables and from dust swallowed in mucus), inhalation (air and dusts) and by dermal contact (soil, air and water). In severe cases, with highly corrosive chemicals, these exposure routes can result in localized damage to the cells of the mouth, trachea and digestive system, nostrils and respiratory system, the skin and the sensitive eyes. However, in most cases, the toxic effects only occur after the pollutant has entered the bloodstream following absorption through either the gut, the lungs or the skin. Once in the bloodstream the chemicals are circulated around the body and undergo metabolism, usually in the liver, or are stored in various organs. Some of the products of this metabolism may be excreted via the kidneys in urine, the digestive tract in faeces, the lungs in exhaled air or sweat from the skin (Rodricks, 1992).

In humans and higher animals metabolic conversion of compounds not essential for normal biological functions takes place mainly in the liver but some metabolism can occur in the lungs, intestines, kidneys and the skin. These conversions are usually catalysed by enzymes but possession of the right enzymes depends on the similarity of the pollutant to commonly encountered substances and evolutionary adaptations. New xenobiotic compounds may remain unaltered or only very slowly metabolized due to the lack of previous exposure to the chemicals. An example of a metabolic process which reduces the toxicity of a pollutant is the conversion of toluene, a neurotoxin which is absorbed through the lungs, into benzoic acid which is much less toxic and more easily excreted than toluene (Rodricks, 1992) (see section 6.2.4 and Figure 3.1).

Toxicology is the study of the effects of poisonous substances on living organisms, including the way in which they gain entry into the organisms. Above a certain concentration, the toxicant has detrimental effects on some biological function. The concentration at which a significant detrimental effect occurs is determined by the *dose response*. The critical (or

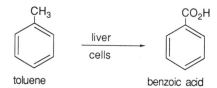

Figure 3.1   The transformation of toluene to less toxic benzoic acid in liver cells (from Rodricks (1992)).

threshold) dose at which toxicity occurs differs between species, sexes and individuals within a species due to genetic and other factors such as the composition of the diet and some illnesses (Manahan, 1991; BMA, 1991) (see section 5.3 for dose response curves of metals).

The *dose*, or degree of exposure of an organism to a toxicant, can be expressed as:

- the amount of toxicant present in the organism (units of mass of toxicant per unit weight of body mass of the organism);
- the amount of the toxicant entering the organism (in the diet, drinking water, or inhaled air in animals and absorbed through the roots or through the leaf cuticle in the case of plants);
- the concentration in the environment of the organism (duration of exposure is important).

The effect of the toxicant dose is called the *response*, and this can vary from no discernible effect to death. Toxicity is commonly categorized on the basis of the duration of exposure, that is acute, chronic and sub-chronic. Acute exposure involves a single dose whereas chronic exposure refers to exposure over a long time period (often almost a lifetime – two years in the case of test rodents). Sub-chronic exposure is dosing over a shorter time period – fraction of a lifetime, such as one eighth of an experimental rodent's lifetime (Rodricks, 1992).

Acute toxicity is caused by fast poisons which include both synthetic and naturally occurring compounds, such as the botulinum toxins produced by the soil bacterium *Clostridium botulinum*, the venom of certain snakes (e.g. rattle snake and cobra) or species of spider (e.g. black widow spider) and plant-derived toxins such as strychnine and nicotine and some synthetic chemicals, including organo-phosphorus compounds (see section 6.5), phosphine (PH_3), phosgene ($COCl_2$) and sodium fluoracetate. These substances are classed as supertoxins because they cause lethal effects in humans at doses of less than 5 mg/kg body weight (as shown in Table 3.1). In general, the toxicity of any chemical depends on its absorption, distribution, metabolism and excretion (ADME) (Rodricks, 1992).

Acute toxic effects are quantified by controlled experiments to determine the dose causing the immediate death of 50% of the organisms

Table 3.1 A classification of toxins on the basis of lethal doses for humans (Rodricks, 1992 and p. 238)

Toxicity rating	Probable lethal dose for humans (mg/kg/body wt)
Practically non-toxic	>15 000
Slightly toxic	5000–15 000
Moderately toxic	500–5000
Very toxic	50–500
Extremely toxic	5–50
Supertoxic	<5

Table 3.2 Lethal doses (LD_{50}) of TCDD (dioxin) for different animal species (from BMA *Hazardous Waste and Human Health* (1991), by permission of Oxford University Press; see section 6.4)

Animal	LD_{50} (μg/kg/body wt)
Guinea pig	1.0
Rat (male)	22
Rat (female)	45
Monkey	>77
Rabbit	115
Mouse	114
Dog	>300
Bullfrog	>500
Hamster	5000

exposed (LD_{50} value). Estimates for humans are extrapolated from values for small mammals (and are subject to many possible errors, such as marked gentoypic variations in susceptibility to a toxicant as shown for TCDD in Table 3.2). The subchronic effects of chemicals are determined by investigating the biochemical and other changes which take place over a period of months. Investigations on chronic effects will examine effects on the lifespan of the organism, cancer induction, changes in geriatric conditions and effects on the offspring caused by exposure of the parent to toxic chemicals. However, it is important to note that the acute and sub-acute effects of toxins determined in laboratory experiments cannot always be relied on to predict responses to the same chemicals in the environment. This is due to intractions between pollutants (antagonistic and synergistic effects) and reactions of the toxicants with the components of the environment (such as adsorption, photodecomposition, acidification and dissolution). In the case of non-carcinogens, it is possible that there may be safe or threshold dose levels of toxins (NOAEL = no observed adverse effect level). However, carcinogens are not considered to have safe or threshold concentrations because a single genetic change may lead to an uncontrolled reaction. Nevertheless, carcinogens differ considerably in potency (e.g. Aflatoxin B_1 is a million times more potent than trichloroethylene).

Space does not permit a detailed discussion of the effects of pollutants on animals and plants and so the following brief lists of effects are provided. The reader is recommended to consult specialized toxicological texts for more details of the subject.

3.2.1 Effects of pollutants on humans and other mammals

The types of response to toxicants which occur in humans and other mammals include (Manahan, 1991):

- alterations in the vital signs of temperature, pulse rate, respiratory rate and blood pressure
- abnormal skin colour
- unnatural odours
- effects on the eye, which include:
    miosis (excessive contraction of pupil)
    mydriasis (excessive pupil dilation)
    conjunctivitis (inflamation of the membrane covering the front of the eyeball)
    nystagmus (involuntary movement of the eyeballs)
- gastrointestinal effects: pain, vomiting, paralytic ileus (stoppage of normal peristalsis)
- central nervous system effects: convulsions, paralysis, hallucinations, ataxia and coma

Sub-clinical (recondite) effects of toxicants in humans and other mammals include (Manahan, 1991):

- damage to the immune system
- chromosomal abnormalities
- modification of the functions of liver enzymes
- slowing of the conduction of nervous impulses

The major types of biochemical effects of pollutants on animals are (Manahan, 1991):

- impairment of enzyme function by the binding of the toxicant to enzymes, coenzymes, metal activators, or enzyme substrates
- alteration of cell membrane or carriers in cell membranes
- interference with lipid metabolism, resulting in excess lipid accumulation
- interference with respiration
- interference with carbohydrate metabolism
- stopping or interfering with protein biosynthesis through toxic effects on DNA
- interference with regulatory processes mediated by hormones or enzymes

3.2.2 Teratogenesis, mutagenesis, carcinogenesis, and immune system defects

Teratogenesis is the creation of birth defects arising from damage to embryonic or foetal cells or from mutations in egg or sperm cells. The biochemical mechanisms of teratogenesis include: enzyme inhibition by xenobiotics, deprivation of essential nutrients and alteration of the placental membrane.

Mutagenesis is the creation of mutations by chemicals or ionizing radiation which bring about alterations to DNA to produce inheritable traits. Although mutations can occur naturally in the absence of xenobiotic substances, most mutations are harmful. Mechanisms of mutagenicity are similar to those of carcinogenesis and teratogenesis.

Chemical carcinogenesis (or cancer) occurs when xenobiotic substances cause uncontrolled cell replication (i.e. cancer). From the public's viewpoint, carcinogenesis is the most commonly associated toxic effect of hazardous substances. Tumours may be benign (when they are contained within their own boundaries) or malignant (when they undergo metastasis, i.e. they break apart and invade other parts of the body) (Masters, 1991).

There are two major steps by which xenobiotic substances can cause cancers: the *initiation stage* and the *promotional stage*. Many chemical carcinogens have the ability to form covalent bonds with DNA, thus altering the DNA so that cell replication becomes uncontrolled. Many chemical carcinogens are either alkylating agents, which attach alkyl (e.g. CH_3) groups, or arylating agents which attach aryl moities (such as phenols) to DNA through the nitrogen and oxygen atoms in the nitrogenous bases (pyrimidine and purine) of DNA (Manahan, 1991).

Chemicals that cause cancer directly are called *primary carcinogens*; however, most xenobiotics involved in carcinogenesis are *pre- or pro-carcinogens* which undergo Phase 1 reactions (lipophilic xenobiotic species undergo enzyme catalysed reactions involving attachment of polar groups such as OH to become more water soluble) or Phase II reactions (in which the polar functional groups in Phase 1 provide sites for conjugation reactions) to produce *ultimate* carcinogens. Vinyl chloride ($CH_2=CHC1$) is a primary carcinogen which can cause a rare form of liver cancer in humans, especially those employed in PVC manufacture (see section 6.4).

The Ames test is the most frequently used procedure to test for mutagenic properties of chemicals. The Ames test is based on the use of liver enzymes to convert potential procarcinogens to ultimate carcinogens. Histidine-requiring *Salmonella* bacteria are inocculated onto a medium that does not contain histidine. Procarcinogen chemicals will increase the chances of mutations occurring which will give rise to forms of the bacteria which can synthesize histidine (Manahan, 1991).

The immune system acts to protect the body from xenobiotic chemicals, infectious agents (viruses and bacteria) and neoplastic cells which give rise

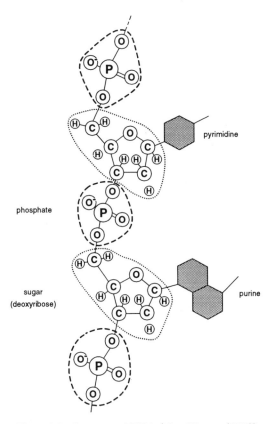

Figure 3.2 Structure of DNA (after Watson (1970)).

to cancerous tissue. Adverse effects on the body's immune system are being increasingly recognized as a consequence of exposure to hazardous substances, UV radiation, etc.

Another reaction is allergy or hypersensitivity when the immune system over-reacts to foreign agents or their metabolites in a self-destructive manner, e.g. Be, Cr, Ni, formaldehyde, pesticides, resins and plasticisers (Manahan, 1991).

Of the ten most important causes of cancer in the US listed in Table 3.3, six of the causes are related in some way to environmental pollution. In addition, the occupational and consumer exposure to chemicals (equal first and equal fourth causes, respectively) will generally involve substances which are also important environmental pollutants. The distinction between these two causes is that the chemicals have not necessarily undergone emission and transport in the environment, although there is a strong likelihood that their manufacture, and/or disposal, may have led to some environmental pollution.

However, Ames and Gold (1990) in an article entitled 'Misconceptions

Table 3.3 The ten most important causes of cancer risk in the USA (USEPA 1986, in Masters, 1991)

Rank	Cause of cancer
1 (=)	Occupational exposure to chemicals (c. 20 000 substances)
1 (=)	Indoor radon (< 20 000 lung cancers/year from exposure in home in the (US)
3	Pesticide residues in foods (6000 cancers/year)
4 (=)	Indoor pollutants (non-radon) – mostly due to tobacco smoke (< 6500 cancers/year)
4 (=)	Consumer exposure to chemicals (cleaning fluids, pesticides, particleboard and asbestos-containing products) (c. 10 000 chemicals)
6	Hazardous/toxic air pollutants
7	Depletion of stratospheric ozone – UVB radiation (caused by pollutants, e.g. CFCs)
8	Hazardous waste sites (inactive) c. 25 000 sites in US
9	Drinking water – radon and trihalomethanes (from chlorination)
10	Application of pesticides (e.g. agricultural workers: high individual risk)

on pollution and the causes of cancer' stress that environmental and dietary exposure to most pollutants, especially synthetic pesticides which are often suspected of being linked to the onset of cancer, is less important in causing cancer than many other factors.

3.2.3 Ecotoxicology

Ecotoxicology is not just concerned with the effects of toxicants on one species but on a wide range of interacting species present in an ecosystem. Behavioural effects, therefore, will be an important consideration in addition to biochemical and physiological responses. Pollutants can affect living organisms in two ways: (i) by being directly toxic or (ii) by causing an adverse change in the organism's habitat which then adversely affects the organism (Smith, 1986). Directly, toxic pollutants first gain entry into an organism and follow an internal pathway which either results in excretion or the disruption of a biochemical process and the onset of toxicity. Indirectly, toxic pollutants alter the physical or chemical conditions of the environment to such an extent that the survival of the organisms is threatened. For example, the depletion of the dissolved O_2 content of a stream, which occurs with the discharge of untreated sewage of silage effluent, can kill many organisms by asphyxiation.

Direct sublethal effects of pollutants also vary between organisms in an ecosystem and some examples of these in animals and plants include (Connell and Miller, 1984):

- Physiological effects on: metabolism, photosynthesis and respiration, osmoregulaton, feeding and nutrition, heartbeat rate, blood circulation, body temperature, water balance.

- Behavioural responses: (individuals) sensory capacity, rhythmic activities, motor activity, motivation and learning (groups and between individuals) migrations, intraspecific attraction, aggression, predation, vulnerability, mating.
- Effects on reproduction: viability of eggs and sperm, breeding and mating behaviour, fertilization and fertility, survival of offspring.
- Genetic effects: chromosome damage, mutagenic and teratogenic effects.
- Effects on growth: body and organ weights, developmental stages.
- Histopathological effects: abnormal growths, membrane abnormalities (respiratory and sensory), reproductive organs.

The organisms within an ecosystem will differ markedly in their tolerance to both direct toxicity and the indirect effects of pollutants. Differences in susceptibility to direct toxicants are clearly illustrated in the LD_{50} data for the lethal dose of dioxin (TCDD) shown in Table 3.2. These data show that not only do species differ in susceptibility but that even males and females of the same species vary, the LD_{50} for the female rat being twice as high as that for males.

**3.3
Assessment of toxicity
risks**

Risk assessment of environmental pollutants is concerned firstly with identifying the hazards which the substances or energy may pose to the health of people, animals and plants, and any damage which they may cause to structures and commodities and, secondly, with estimating the probability of these types of harm occurring. It is important to know the pathways and dose-response relationships for all hazardous substances and possible target organisms (including humans, farm livestock, crop and other plants as well as surrounding ecosystems). Much of this information is now available for most of the commonly encountered pollutants although it is constantly being added to and improved. Nevertheless, it is important to recognize that there is a great deal of uncertainty inherent in the characterization of risk.

3.3.1 Pollutants in contaminated land

Taking contaminated land as an example, Table 3.4 summarizes the commonly encountered types of risk associated with the pollutants.

Empirical transfer coefficients have been determined and models have been developed for quantifying dose–response relationships. In practice, many states or countries have established lists of critical concentrations (or 'trigger concentrations') for the risk assessment of site and environmental survey data. The basis for these different sets of values varies according to the target groups they are intended to protect from the effects of the pollutants.

Table 3.4 Examples of the risks from chemicals in contaminated land (adapted from Alloway (1992) based on Beckett and Sims (1986) and ICRCL (1987))

Human and animal toxicity and carcinogenicity
- the direct ingestion of contaminated soil (mainly by children and grazing livestock) e.g. CN, As, Pb, PAHs
- inhalation of dusts, toxic gases and vapours from the contaminated soil, e.g. benzene, solvents, Hg, CO, HCN, H_2S, PH_3 and asbestos
- uptake by plants of contaminants hazardous to animals and people through the food chain, e.g. Cd, As, Pb, Tl, PAHs
- contamination of drinking water supplies, e.g. phenols, CN, SO_4, soluble metals, pesticides, e.g. atrazine in groundwater and permeation of water pipes by solvents
- skin contact, e.g. tars, phenols, asbestos, radionuclides, PAHs, PCBs and PCDDs

Phytotoxicity
- SO_4^{2-}, B, Cu, Ni, Zn, herbicide residues

Fire and explosion
- CH_4 and high calorific wastes from landfills
- coal dust
- petroleum, solvents

Deterioration of building materials and services
- SO_4^{2-}, SO_3^{2-}, Cl^-, coal tar, phenols, mineral oils, solvents

---

Examples of tables of critical (trigger) concentrations for organic pollutants used in different countries are given below (the equivalent data for heavy metals is given in section 5.3.6):

(a) The Netherlands previously had a scheme which gave three indicative values for a wide range of pollutants: A, the 'normal' reference value; B, the value at which it is necessary to conduct further investigations into the form and bioavailability of the pollutant; and C, the intervention value above which the soil definitely needs cleaning-up (Moen *et al.*, 1986). This system was superseded by an effect-oriented scheme of 'Environmental Quality Standards for Soil and Water' (Netherlands Directorate General for Environmental Protection, 1991). These standards are based on ecological function and comprise 'target values' (TV) for soils and waters which represent the final environmental quality goals for the Netherlands. In the case of surface and groundwaters both target and limit values are given. The limit value is equivalent to the maximum permissible risk level and is intended to indicate the environmental quality to be achieved in a given period. It is intended that limit values would be progressively reduced until they reach the target value. However, the authors of the standards recognized that pollutants generally have long residence times in soils and so they have not given limit values for soils, only target values. The intervention values (*C* values) of the 1986 scheme still remain because if the concentrations of a pollutant reached this level there would definitely be a need for a thorough investigation and remedial action. The target values are given for a 'standard soil' with 10% organic matter and 25% clay, but formulae are provided to allow the values to be calculated for soils with a wide range of clay and organic matter contents. For example,

Table 3.5   Guide values (in μg/g) and quality standards used in the Netherlands for assessing soil contamination by organic and inorganic substances. See sections 6.2 for PAHs, 6.3 for benzene, 6.5 for PCBs, PCDDs and PCDF and chapter 4 for PCP.

Category	A	B	C	TV*
Inorganic pollutants				
CN (total free)	1	10	100	1
CN (total complex)	5	50	500	5
Br	20	50	300	
S	2	20	200	
Polycyclic aromatics				
PAHs (total)	1	20	200	
Naphthalene	0.1	5	50	15
Anthracene	0.1	10	100	50
Benzo-[a]-pyrene	0.05	1	10	25
Chlorinated hydrocarbons				
CH total	0.05	1	10	
PCBs	0.05	1	10	
Chlorophenols (total)	0.01	1	10	
Pentachlorophenol	—	—	—	2
Pesticides				
Pesticides (total)	0.1	2	20	
Aromatic compounds				
Aromatics (total)	0.1	7	70	
Benzene	0.01	0.5	5	
Toluene	0.05	3	30	
Phenols	0.02	1	10	
Other organic compounds				
Cyclohexane	0.1	5	60	
Pyridine	0.1	2	20	
Gasoline	20	100	800	
Mineral oil	100	1000	5000	

(A= reference value, B = test requirements, C = intervention value, from 1986 scheme; *TV = target value in 1991 Environmental Quality Standards for Soils and Waters)

the Pb target value is determined by the formula: $50 + L + H$ (where L = % clay and H = % organic matter).

Examples of reference values from the Netherlands' Environmental Quality Standards for Soil and Water are given in Table 3.5 (data for metals in both soils and waters in this Netherlands list are given in chapter 5).

(b) In the United Kingdom the Department of the Environment set up an Interdepartmental Committee for the Redevelopment of Contaminated Land (ICRCL) to draw up list of trigger concentrations for contaminants (DOE, 1987). These trigger concentrations are more pragmatic than the Netherlands Environmental Quality Standards and are based mainly on the risk to human health. Unlike the Dutch standards, the ICRCL values vary for different proposed uses of the contaminated land. The lowest values are given for garden soils where vegetables are likely to be grown,

Table 3.6 UK Department of the Environment (ICRCL) trigger concentrations for contaminants associated with former coal carbonization sites.

| Contaminant | Proposed use | Trigger concentrations (μg/g) | |
		Threshold	Action
PAHs	Gardens, allotments	50	500
	Landscaped areas	1000	10 000
Coal tar	Gardens, allotments	200	—
	Landscaped areas, openspace, buildings, hard cover	500	—
Phenols	Gardens, allotments	5	200
	Landscaped areas	5	1000
Free cyanide	Gardens, allotments	25	500
	Buildings, hard cover	100	500
Complex	Gardens, allotments	250	1000
cyanides	Landscaped areas	250	5000
	Buildings, hard cover	250	NL
Thiocyanate	All uses	50	NL
Sulphate	Gardens, allotments	2000	10 000
	Landscaped areas, buildings	2000	50 000
	Hard cover	2000	NL
Sulphide	All uses	250	1000
Sulphur	All uses	500	20 000
Acidity	Gardens, etc.	pH < 5	pH < 3

NL = no limit set because contaminant does not pose a particular hazard when land used for this purpose.

with higher values for parks and open spaces, and the highest values for land to be developed for industrial uses where the transfer of pollutants from the soil to plants is not likely to be significant in terms of its impact on human health. Examples of the ICRCL values for both general sources of pollution and more specifically for land affected by coal carbonization (gas and coke works) are given in Table 3.6.

(c) In Canada, the National Contaminated Sites Remediation Program published interim Environmental Quality Criteria in 1991 for use in the evaluation of contaminated sites (Table 3.7). The values given are intended to enable sites to be classified as high, medium or low risk according to their impact (current or potential) on human health and ecosystems. It is a screening system and is not intended to be a quantitative risk assessment for individual sites.

The 1991 Environmental Quality Standards for the Netherlands (Table 3.5) give target values for soils which will be very difficult and expensive to achieve, especially in the case of some of the ubiquitous contaminants, such as Pb. The threshold value for Pb under the ICRCL scheme used in the UK (Table 3.6) is 500 μg/g. This implies that concentrations below this figure should not cause problems. In contrast, the Dutch target value for Pb is 85 μg/g (for a standard soil) and the Canadian background value (benchmark level) is even lower at 25 μg/g but the trigger concentration for

Table 3.7 Selected data from the Interim Canadian Environmental Quality Criteria for Contaminated Sites (values in μg/g) (Canadian Council of Ministers of the Environment, 1991) (Reproduced with permission of the Minister of Supply and Services Canada, 1993.)

(a) Metals

Metal	Background	Agricultural	Residential	Industrial
As	5	20	30	50
Ba	200	750	500	2000
Be	4	4	4	8
Cd	0.5	3	5	20
Cr^{6+}	2.5	8	8	—
Co	10	40	50	300
Cu	30	150	100	500
CN (free)	0.25	0.5	10	100
CN (total)	2.5	5	50	500
Pb	25	375	500	1000
Hg	0.1	0.8	2	10
Mo	2	5	10	40
Ni	20	150	100	500
Se	1	2	3	10
Ag	2	20	20	40
Sn	5	5	50	300
Zn	60	600	500	1500

(a) Organic pollutants

	Background	Agricultural	Residential	Industrial
Monocyclic hydrocarbons				
Benzene	0.05	0.05	0.5	5
Chlorobenzene	0.1	0.1	1	10
Toluene	0.1	0.1	3	30
Phenols				
Phenols (each)	0.1	0.1	1	10
Chlorophenols (each)	0.05	0.05	0.5	5
PAHs				
benzo-[a]-pyrene	0.1	0.1	1	10
naphthalene	0.1	0.1	1	10
Chlorinated hydrocarbons				
Chlorinated aliphatics				
(each)	0.1	0.1	5	50
Chlorobenzenes (each)	0.05	0.05	2	10
Hexachlorobenzene	0.1	0.05	2	10
PCBs	0.1	0.5	5	50
PCDDs and PCDFs	0.00001	0.0001	0.001	—

the remediation of agricultural land is 375 μg/g. The Dutch scheme is based on the ecological effects of contaminants and therefore soils meeting the target values would be suitable for any use, such as food production or nature conservation. On the other hand, the UK and Canadian values take into account different uses of land. The UK value of 500 μg/g Pb is the most achievable, because many urban soils in the UK would come within this range and the more excessively polluted sites would be considered unacceptable.

Table 3.8 Guideline and maximum acceptable concentrations for metal and other inorganic pollutants in drinking waters (WHO (1984); Murley (1992); Canada Council of Ministers of the Environment (1991); Manahan (1991) (µg/l)) (Reproduced with permission of the Minister of Supply and Services Canada, 1993.)

	EC/WHO	Canada*	USA*
As	50(W)	25	50
B	1000	5000	1000
Ba	100	1000	—
Fe	50	<300	50
Mn	20	<50	50
Cd	5(W)	5	10
Cr	50(W)	50	50
CN	100(W)	200	—
Pb	50(W)	10	5
Hg	1(W)	1	—
Cu	<3000 (100 at works)	<1000	1000
U	—	100	—
Zn	<5000 (100 at works)	<5000	5000
F	1500 (8–10°C) 700 (25–30°C)	1500	800–1700 (depending on temperature)

EC = 80/778/EEC Quality of water for human consumption
(W) = WHO 1984 guideline values
* Maximum acceptable values

Of particular relevance is the intended future use of the contaminated land. In some countries, such as the Netherlands, the intention is to ameliorate the site to a specification which will allow the land to be used for any purpose (multifunctionality). The more pragmatic approach in other countries, such as the UK, is to relate the site quality specification to the intended use. For example, a contaminated site required for development as a warehouse complex would not need to have such low concentrations of toxic compounds as a site to be used for housing where the residents may grow vegetables in their gardens. However, any explosive hazard, such as methane release, would need to be removed for both new uses of the site. Even though the use of a site may not require a high degree of clean-up, the remaining contaminants may migrate within the soil to the ground-water, undergo chemical changes or remain a potential problem for future uses of the site.

3.3.2 Pollutants in drinking water

The concentrations of pollutants in drinking water is very important since adults drink on average 1.5 l/day and are therefore likely to be more rapidly affected by pollutants from this route than from the diet. A selection of some of the guideline values currently in use are given in Tables 3.8 and 3.9.

Table 3.9 World Health Organization guideline values for selected micropollutants in drinking waters (WHO, 1984)

Organic pollutant	Guideline value (μg/l)
Aldrin and dieldrin	0.03
Benzene	10
Benzo-[a]-pyrene	0.01
Chlordane (total isomers)	0.3
Chloroform	30
2,4-D (2,4-dichlorophenoxyacetic acid)	100
DDT (total isomers)	1
1,2-Dichloroethane	10
1,1-Dichloroethane	0.3
Heptachlor and heptachlor epoxide	0.1
Hexchlorobenzene	0.01
Gamma-HCH (lindane)	3
Methoxychlor	30
Pentachlorophenol	10
2,4,6-Trichlorophenol	10

The organic pollutants in Table 3.9 are acknowledged to be those closely related to human health. It is interesting to note that for the two isomers of dichloroethane, the 1,1-dichloroethane isomer is considered to be 33 times more hazardous than the 1,2 isomer. The lowest guideline values, which imply a high toxicity hazard, are those for the PAH benzo-[a]-pyrene and hexachlorobenzene.

3.3.3 Toxic or explosive gases and vapours

Although a wide range of toxic gases and fumes can be released in accidents in chemical plants or stores, there are several gases and vapours which are commonly encountered in many pollution situations, including contaminated land, waste disposal and fuel combustion. The safe concentrations for some of these gases and vapours are given in Table 3.10

Table 3.10 Critical concentrations for common gaseous pollutants (% v/v) (Smith, 1991)

Contaminant	Toxicity	Flammability	Trigger for potential hazard
Carbon dioxide	0.5	—	0.125
Carbon monoxide	0.005	12–75	0.00125
Sulphur dioxide	0.0005	—	0.000125
Hydrogen sulphide	0.001	4.3–45.5	0.0025
Methane	14	5–15	0.25
Petrol	0.1	1.4–7.6	0.025

(Smith, 1991). It is interesting to note that the most hazardous of these, when judged by the concentration causing toxicity, is SO_2.

With regard to combustibility, calorific values greater than 10 MJ/kg are normally regarded as a potential combustion hazard but materials with values between 2 and 10 MJ/kg could also pose a problem in some circumstances. Materials with calorific values below 2 MJ/kg are unlikely to burn (Fleming, 1991).

References

Alloway, B. J. (1992) Chapter 5 in Harrison, R. M. (ed) *Understanding Our Environment* (2nd edition), Royal Society of Chemistry, Cambridge.

Ames, B. N. and Gold, L. S. (1990) *Agnew. Chem. Int. Ed. Engl.*, **29**, 1197.

Beckett, M. J. and Sims, D. L. (1986) in Assink, J. W. and van den Brink, W. J. (eds) *Contaminated Soil*, Martinus Nijhof, Dordrecht.

British Medical Association (1991) *Hazardous Waste and Human Health*, Oxford University Press, Oxford.

Canada Council of Ministers of the Environment (1991) *Interim Canadian Environmental Quality Criteria for Contaminated Sites*, Report CCME EPC-CS34, Winnipeg, Manitoba.

Connell, D. W. and Miller, G. J. (1984) *Chemistry and Ecotoxicology of Pollution*, John Wiley and Sons, New York.

Inderdepartmental Committee on the Redevelopment of Contaminated Land (1987) *Guidance on the Assessment and Redevelopment of Contaminated Land*, Department of the Environment, London.

Fleming, G. (1991) Chapter 1 in Fleming, G. (ed) *Recycling Derelict Land*, Thomas Telford, London.

Manahan, S. E. (1991) *Environmental Chemistry* (5th edition) Lewis Publishers, Chelsea, Mich.

Masters, G. M. (1991) *Introduction to Environmental Engineering and Science*, Prentice Hall, Englewood Cliffs, NJ.

Ministry of Housing, Physical Planning and Environment, Directorate General Of Environmental Protection (1991) *Environmental Quality Standards For Soil and Water*, Ministry of Housing, Physical Planning and Environment, Leidschendam, Netherlands.

Moen, J. E. T., Cornet, J. P. and Evers, C. W. A. (1986) in Assink, J. W. and van den Brink, W. J. (eds) *Contaminated Soil*, Martinus Nijhoff, Dordrecht.

Murley, L. (ed) (1992) *1992 Pollution Handbook*, National Society for Clean Air and Environmental Pollution, Brighton.

Rodricks, J. V. (1992) *Calculated Risks*, Cambridge University Press, Cambridge.

Smith, S. (1986) Chapter 5 in Hester, R. E. (ed) *Understanding Our Environment* (1st edition), Royal Society of Chemistry, London.

Smith, M. (1991) Chapter 5 in Fleming, G. (ed) *Recycling Derelict Land*, Thomas Telford, London.

Smith, S. (1992) Chapter 8 in Harrison, R. M. (ed) *Understanding Our Environment* (2nd edition), Royal Society of Chemistry, Cambridge.

Watson, J. D. (1970) *The Double Helix*, Weidenfeld & Nicholson, London.

World Health Organisation (1984) *Guidelines for Drinking Water Quality*, Vol 1, WHO, Geneva.

Analysis and monitoring of pollutants – organic compounds 4

4.1
Chromatography

Chromatography is the dominant analytical technique for the identification and quantification of organic pollutants. Classical chromatography was first used by Tswett in 1906 for the analysis of plant extracts which were charged on to the top of a column of chalk and eluted with a downward flow of petrol. Chromatograms of this sort depend on the different affinities of the components of a mixture for the stationary phase; this could be chalk, alumina, kieselguhr or other surface active solid. Partition into the downflowing solvent then carries the separated compounds to the end of the column in sequence. There are a number of practical details which are essential for efficiency:

1. The charge should be concentrated at the column head since ideally it would be located as a point source.
2. Sufficient solid phase must be provided so as to ensure that it is not overloaded with material.
3. Care must be taken in choosing the solvent which must have sufficient strength or affinity with the charge to ensure that this becomes desorbed from the stationary phase. If too polar a solvent is chosen it will limit interaction with the solid phase which will no longer differentiate between one component and another; that is why they will not be resolved.
4. The flow rate must be on a reasonable time scale, but not so fast as to detract from the efficient 'mass transfer' of components between the stationary solid phase and the flowing solvent (see p. 69).

Modern chromatography has many specialized applications including paper for the separation of amino acids and peptides and ion-exchange for the analysis of mixtures of metal ions. Gel chromatography is based on the principle of the molecular sieve which retains smaller molecules and excludes larger molecules which pass rapidly through the system. For our purpose it is necessary to concentrate on those methods of most use for the analysis of pollution problems.

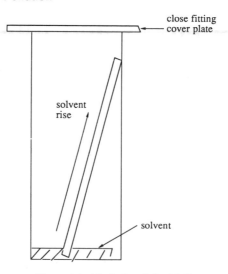

Figure 4.1 Typical tank for TLC.

4.2
Thin layer
chromatography (TLC)

This was first used in 1938 by Izmailov and Shraiber. It is similar in application to paper chromatography and has replaced it to a considerable extent. The adsorbent or stationary phase is spread evenly on a plate of glass or a sheet of metal foil. For analytical purposes the layer will be about 0.1 mm thick and will only accept a small point charge if the adsorbent is not to be saturated. For chromatography of larger amounts, layers 1–2 mm thick can be used and they can be charged as a basal streak at the rate of about 1 mg/cm length.

The plate is usually developed by standing it in a reservoir containing a small quantity of solvent or mixture of solvents (Figure 4.1).

The tank must be covered to ensure that the space around the plate is saturated with vapour and under these conditions the solvent will rise by capillary action to a height of about 15 cm. This will be complete in about 20 minutes and the speed and simplicity of the technique accounts for its widespread application. One important use is in monitoring the eluent from larger-scale column chromatograms and another is in evaluating solvents for use in high pressure liquid chromatography (HPLC).

The adsorbent most used is silica gel, but alumina, cellulose and ion exchange resins can also be applied. The range of solvents is extensive and accounts in considerable part for the power of the method.

Detection.   Unlike paper this presents no problem. Colourless substances of low reactivity can be visualized on silica and alumina by spraying the developed plate with a 10% solution of concentrated sulphuric acid in ethanol and leaving in an oven until oxidized traces appear. Other strong

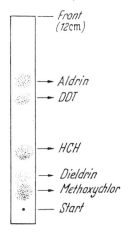

Figure 4.2   TLC of pesticides.

oxidizing agents such as permanganate can be used. Most often prepared plates are used which include fluorescent markers and they reveal the eluted spots on placing under a UV lamp. The spots appear as darker zones against a green (254 nm) or purple (366 nm) background.

4.2.1   Separation of pesticides

Figure 4.2 shows the end result of eluting a mixture of five organochlorine pesticides on silica gel (Stahl, 1969). The solvent chosen was cyclohexane 80%:chloroform 20%. A suitable alternative would have been the use of *n*-hexane with alumina as the stationary phase. The spots show up clearly with a fluorescent marker and may be identified by running a wider plate on which the individual reference compounds are spotted out separately along the base. Alternatively, the spots can be scraped off and extracted with a hot solvent and the extract identified by mass spectrometry. The technique works well for this group of complex individual compounds; only for dieldrin and methoxychlor is separation incomplete.

4.2.2   Separation of metal cations

Figure 4.3 is a block diagram which represents the movement of a group of heavy metal ions dissolved as their nitrates using *n*-butanol containing 1.5 M hydrochloric acid (15%) as solvent and acetylacetone (0.5%) for complexation. The record shows the mixture running on the right and individual ions which were spotted on for identification by reference. After development the spots are detected by drying, exposure to ammonia and

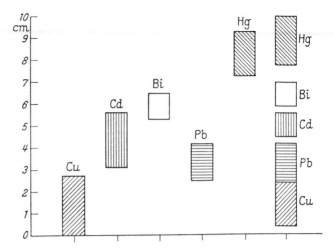

Figure 4.3 TLC of heavy metal ions.

then to hydrogen sulphide. The metals are revealed as their black or brown sulphides, save for cadmium which is yellow. Metal anions, for example CrO_4^{-2} and AsO_3^{-3}, can also be run on TLC with eluents such as acetone:water or alcohol:water and after extraction from the plate their identity may be confirmed by atomic absorption (AA) spectroscopy.

4.3
Gas liquid
chromatography (GLC)
(Braithwaite, 1985)

For the environmental scientist this procedure is of first importance. It was originally developed in 1952 by Martin and James and as the name implies it depends on partition between a flowing gaseous phase and a stationary liquid phase. Although more sophisticated than TLC, the same provisions must be met if the column is to work efficiently.

Figure 4.4 shows a glass column up to 4 metres long with an internal diameter (i.d.) of 2–4 mm. It is usually in the form of a spiral so that it will fit conveniently within the oven. The entry port is adapted to connect a

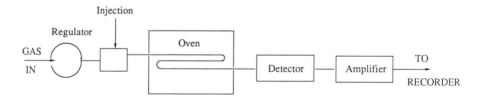

Figure 4.4 Diagram of a GLC column and essential hardware.

flow of inert gas, nitrogen or, better, helium, which has passed through a flow control and a sieve to remove traces of moisture. The entry port also includes a re-sealing silicone rubber septum cap, so that the sample may be injected through it into the material at the beginning of the column; this ensures that the analyte is initially concentrated as the narrow band essential for efficient separation.

The molecular weight of the analyte is restricted to about 500 Da and if it is of low volatility it must be run at a temperature high enough to ensure that it has a significant vapour pressure. The containing oven includes a fan to give as even a temperature distribution as possible. For practical purposes the maximum operating temperature is about 260°C and the injection port must be maintained 50°C above this to prevent condensation of the sample.

Clearly attention must be paid to the nature of the analyte. Volatile hydrocarbons are readily examined but polar substances like alcohols, amines and phenols may require derivatization. Carboxylic acids must be esterified before the analysis. In this context silylation is important; here a polar substance is converted to a silyl ether derivative by reaction in a volatile solvent with trimethylchlorosilane (equation 4.1) or hexamethyldisilazane (equation 4.2):

$$Ph{-}OH \ + \ Cl - SiMe_3 \quad \rightarrow Ph - O - SiMe_3 \qquad + \ HCl \qquad (4.1)$$

phenol   trimethylsilyl   phenyl trimethylsilyl
                     chloride   ether

$$2\,Ph{-}NH_2 + Me_3Si{-}N{=}N{-}SiMe_3 \ \rightarrow \ 2\,PhNH{-}SiMe_3 + N_2 \qquad (4.2)$$

aniline   trimethylsilyl aniline

These reactions take only a few minutes for completion at room temperature; the solvent is then evaporated in a stream of nitrogen and replaced with that chosen for injection onto the column. Silylation is also an essential pretreatment for the column before it is packed to ensure that 'hot' spots are blocked and do not spoil the separation.

Packed columns require a solid particulate support that is inert and has a large surface area/unit volume. Suitable materials may be firebrick derived, such as Chromosorb P or diatomaceous substances derived from filter aids. They need to be acid-washed and silylated before the stationary phase is applied to them as a dilute solution in a volatile solvent. After removal of the solvent the surface saturated material is packed into the column; Chromosorb W has pores of 8 μm diameter, a surface of $1m^2$/g and it can be packed at a density of 0.24 g/cm^3.

The diagram shows how, as the gas flow sweeps the charge along the column, a succession of equilibria is established between analyte in the stationary phase and that in the gaseous phase. Given that the components have different affinities for the stationary phase, those with the greater affinity will lag behind and eventually separate so that a volume of pure

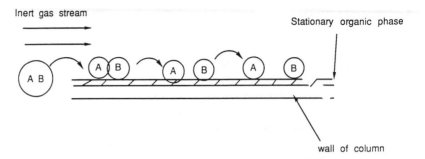

Figure 4.5   Diagrammatic impression of the progress of separation.

carrier gas lies between them. The process bears some relation to fractional distillation whose efficiency was measured in theoretical plates, that is the number of points in the column where equilibrium was established between ascending vapour and the condensed liquid.

The enhanced efficiency of GLC arises because the stationary phase can be chosen so as to bias the partition in favour of particular compounds. Here use is made of the activity term which will be close to unity when the substrate is similar to the stationary phase, but will diverge if the substance differs in type. This is true for differences in polarity and Table 4.1 gives examples. With this low polarity hydrocarbon as the stationary phase the alkanes are eluted in the order of their boiling points as shown down the table. However, considerable divergence is seen for the polar carbonyl compound and still more for the hydrogen-bonded propanol.

An example of how this may be put to practical use is in the separation of benzene (b.p. 80.1°C) from cyclohexane (b.p. 81.4°C). This is a near impossibility by fractionation but is straightforward by GLC with the high boiling ester dinonyl phthalate (**1**) since the common aromatic structure favours benzene and hence its activity towards this phase diverges from that of cyclohexane.

Table 4.1   Activity coefficients in n-hexadecane ($C_{16}H_{32}$) at 30°

Compound	Coefficient	b.p.(°C)
1. n-Pentane	0.88	36
2. n-Hexane	0.89	69
3. Cyclohexane	0.77	81
4. Propionaldehyde (C_2H_5CHO)	4.0	49
5. Acetone (CH_3COCH_3)	6.3	56
6. n-propanol	31.5	98

Relative retention.   In the above example the retention volume for benzene (V_a) is the volume of carrier gas which passes before benzene is eluted from the head of the column. Similarly the retention volume of cyclohexane is (V_b). In practice the ratio $V_a/V_b = 1.6$, very different from the b.p. ratio clearly showing how benzene is retained by the phase with a common structural element. The nonyl phthalate would evidently be useful for the separation of carboxylic esters.

Polar substances such as alcohols and amines are best chromatographed on polar phases such as polyethylene glycol (**2**) and ethylene glycol succinate (**3**).

4.3.1   Detection of eluted substances

The flame ionization detector (FID).   This is widely used, extremely sensitive and depends on the combustion of eluted substances within a small flame of hydrogen burning in an air stream at the head of the column (Figure 4.6).

When eluted carbon compounds burn, free radicals are released which combine with oxygen, releasing electrons to produce an ion current. This is proportional to the amount of eluant and may be amplified and used to drive a pen recorder. To be significant the visual response has to be at least twice that of recorder background noise. The FID is extremely sensitive and responds to a concentration of one picogram $(10^{-12}g)$/ml of carrier gas; the limit of detection is about a microgram. This will vary with the analyte: hydrocarbons respond more strongly than alcohols and other compounds with electronegative substituents.

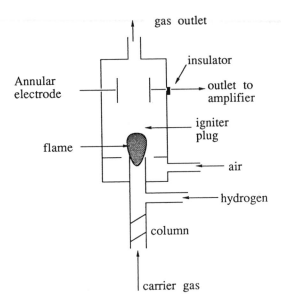

gas outlet

insulator

Annular
electrode

outlet to
amplifier

igniter
plug

flame

air

hydrogen

column

carrier gas

Figure 4.6 Diagram of a flame ionization detector.

The response varies with individual compounds and if a quantitative result is needed a calibration plot must be made setting out detector response over a range of concentrations of each component. Alternatively the chromatogram can be re-run after 'spiking' the sample with a known aliquot and comparing the peak areas obtained in the two instances. A superior method is the introduction of a known quantity of substance similar to the analyte to act as an internal standard and this is illustrated in the example of Figure 4.24.

The electron capture detector (ECD).   The sensitivity of the FID falls away for compounds like the PCBs (p. 232) containing electronegative atoms because they do not burn well (note that the polybromobiphenyls are used as fire retardants). The FID response is also poor for polluting gases such as CO, CO_2 and SO_2. Very fortunately it is just this type of electronegative character which enhances the response of the ECD (Figure 4.7).

The detector is set up with a steady electron current following from a radioactive source, usually a piece of [63]Ni foil. When molecules with an affinity for electrons enter the detector space some are captured and the current falls and this imbalance can be amplified and recorded.

The sensitivity of the ECD can be two orders of magnitude better than the FID and in use care must be taken not to inject excessive amounts of the analyte because this will overload the detector. If a solvent such as chloroform or carbon tetrachloride is used to extract the sample it must

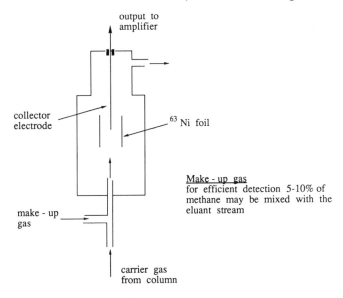

Figure 4.7 Diagram of an electron capture detector.

be completely removed in a stream of inert gas (nitrogen) and replaced by a hydrocarbon before making the injection. The response of the ECD to $CHCl_3$ and CCl_4 is about 10^4 times that of a monochloroalkane.

4.3.2 Principal parameters

Figure 4.8 shows a typical pen recorded gas chromatogram of three components. It is clear that a satisfactory separation has been achieved, but note that the peak with the longest retention time (t_r) or with the largest elution volume (flow rate $\times$ t_r) is appreciably broadened. This is the direct result of the progress of partition between phases; each transfer, or each plate, broadens the sample distribution and as a result the peaks achieve the shape of a Gaussian curve. If the width at the base (w) results from n transfers (plates) then:

$$n = 16 \ (t_r/w)^2$$

or for width at half-height

$$n = 5.54 \ (t_r/w)^2$$

For a column of length L cm: L/n = HETP, the height equivalent to a theoretical plate. For an efficient column this will be small and of the order of 0.05 cm or 2000/m.

When the analyte contains a range of substances of varying polarity and

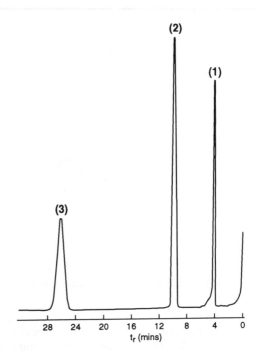

Figure 4.8   Typical pen-recorded gas chromatogram (Gopalan, 1977).

molecular weight some will inevitably be broadened by long retention. In these circumstances the record can be improved by temperature programming. Here the chromatogram is run at a temperature consistent with the efficient elution of the more volatile components and thereafter the oven is programmed for a temperature rise at some 2–10°C/min until it reaches an upper temperature range suitable for the separation of the less volatile components. The retention time of the latter and the spread is thereby reduced as can be seen in Figure 4.9. This shows the separation of 3 alkylbenzenes and 17 *n*-alkanes; their thermal stability permits running to a high maximum temperature.

4.3.3   Optimum operating conditions

The factors are defined within the Van Deemter equation:

$$\text{HETP}(H) = A + B/\bar{u} + C.\bar{u} \tag{4.3}$$

where $\bar{u}$ is the mean velocity of the mobile phase; A is the component of

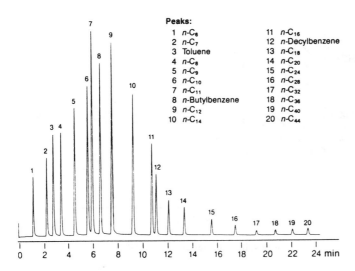

Peaks:

1	$n\text{-}C_6$	11	$n\text{-}C_{16}$
2	$n\text{-}C_7$	12	n-Decylbenzene
3	Toluene	13	$n\text{-}C_{18}$
4	$n\text{-}C_8$	14	$n\text{-}C_{20}$
5	$n\text{-}C_9$	15	$n\text{-}C_{24}$
6	$n\text{-}C_{10}$	16	$n\text{-}C_{28}$
7	$n\text{-}C_{11}$	17	$n\text{-}C_{32}$
8	n-Butylbenzene	18	$n\text{-}C_{36}$
9	$n\text{-}C_{12}$	19	$n\text{-}C_{40}$
10	$n\text{-}C_{14}$	20	$n\text{-}C_{44}$

Figure 4.9   Temperature-programmed separation of a hydrocarbon mixture. 50.8 mm × 0.3 mm Me silicone on Chromosorb column, programmed from −25°C at 15°/min to 350°C. Carrier gas. He at 30 ml/min with FID detection. Total neat sample 0.1 µl (approx. 80 µg) (SUPELCO, 1990).

eddy diffusion; B is the component of longitudinal diffusion; and C is the component of mass transfer.

For efficiency these variables must be chosen so as to minimize H and this introduces an element of compromise.

Eddy diffusion (A).   This is a constant factor which arises from the range of path lengths which the solute molecules can take around the particles of the stationary phase, including some movement into the pores. The best remedy is the use of particles regularly shaped and as small as possible, consistent with pressure being available to maintain the flow through them.

Longitudinal diffusion (B).   Varies with the flow rate; when this is low there is every opportunity for the solute to spread through the gaseous phase from points of high concentration to those of lower concentration. This natural redistribution runs counter to the progress of partition and it can be seen from equation 4.3 that it varies inversely with flow rate.

Mass transfer effect (C).   This is concerned with the resistance to transfer of material from the liquid to the flowing gaseous phase. This can only reach a true equilibrium if the phases are stationary, an unrealizable condition although low flows are favourable. With rising flow rate the two

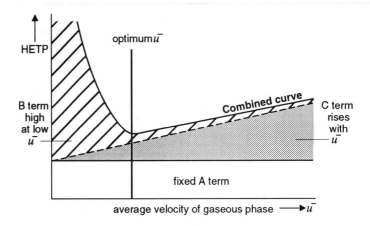

Figure 4.10   The variation in H with flow rate.

phases get further out of step and the $C.\bar{u}$ terms get progressively larger, hence it is clear that to minimize the combined effect of $B/\bar{u}$ and $C.\bar{u}$ an intermediate rate must be established. Figure 4.10 shows the summation of the three factors and the minimum in the curve corresponds to the best compromise. In practice the flow is set slightly above this because even a small fallback leads to a steep rise in H.

4.3.4   Capillary columns for GLC

These were first developed by Golay in 1958 and they now represent the most effective columns available. They are of steel or flexible fused silica with an i.d. of 0.2–0.5 mm and length in the range of 10–50 m. They may be wall coated open tubular (WCOT), which are made by forcing a solution of the stationary phase slowly through while gradually raising the temperature, or support coated open tubular (SCOT), which include a fine layer of support material and can accept rather more charge than the WCOT although this is still limited to about 10 μg.

The HETP of capillary columns is about 0.03 mm or 30×10^3/m which is about 10 times that obtainable with packed columns. Capillary columns have other advantages including:

1. neglible broadening by eddy diffusion;
2. shorter elution times than packed columns;
3. narrow peaks which ensure a low detection limit;
4. low back pressure because of the open end and hence very long columns can be used.

4.3.5   Analysis of urban air pollution

Even allowing for the sensitivity of capillary GLC, the direct measurement of all pollutant levels in air is not possible and so concentration is carried out by pumping a measured volume of air through an adsorptive trap. This will be a column packed with a material such as Tenax (**4**), a polymer of 2,6-diphenyl-*p*-phenylene oxide:

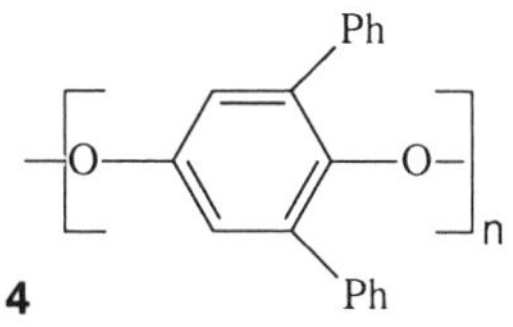

The procedure for desorption is critical and quite complex equipment has been engineered for the purpose. In outline the Tenax column is placed in a preheated oven and the analyte swept out in an inert gas stream into a cooled trap. This trap is then placed in the input line to the GLC column and heated for a few seconds only. This time restriction is necessary because the charge must be as concentrated as possible and prolonged tailing must be avoided, although this means that injection will be incomplete. Figure 4.11 shows the recorded pollutants in air at Bilthoven in May 1990. Notice the intense peaks (1, 2, 3, 6) arising from the most volatile compounds and the significant peaks due to the toxic components benzene (27) and hexane (24). Prolonged sampling of the air may not be possible because the most intense peaks would overload the capillary column. If a quantitative result is required the injection procedure will have to be calibrated by loading known amounts of each component in turn in the Tenax column and evaluating the proportion discharged through the GLC column.

4.3.6   Detection by mass spectrometry (Suelter, 1991)

This is not only a very sensitive recorder of peaks as they are eluted from the column but it can also identify the component.

In its basic construction the mass spectrometer focuses ions produced by bombardment of the parent molecules with a high energy beam of electrons (*c.* 70 ev). A fruitful collision:

$$M + e \rightarrow M\dot{+} + 2e \tag{4.4}$$

dislodges an electron leaving a molecular ion ($M\dot{+}$) which can be focused

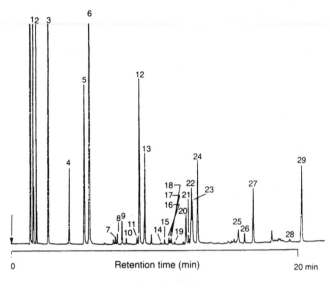

Peaks:

1.	ethane	16.	2-methyl-2-butene
2.	ethene	17.	1-pentene
3.	propane	18.	2-methyl-1-butene
4.	propene	19.	cis-2-pentene
5.	isobutane	20.	methylcyclopentane
6.	n-butane	21.	cyclohexane
7.	trans-2-butene	22.	2-methylpentane
8.	1-butene	23.	3-methylpentane
9.	isobutene	24.	hexane
10.	cis-2-butene	25.	2/3-methylhexane
11.	cyclopentane	26.	n-heptane
12.	2-methylbutane	27.	benzene
13.	n-pentane	28.	n-octane
14.	3-methyl-1-butene	29.	toluene
15.	trans-2-pentene		

Figure 4.11 Hydrocarbons in smog at Bilthoven. 50 m × 0.5 mm fused silica capillary column, temperature programmed at 40°C initially then at 10°C/min to 200°C. Air sample 0.41 and FID detector (CHROMPAK, 1992).

on a slit because the radius (r) of the path it follows within a magnetic field is a function (equation 4.5) of its mass, the field strength (B) and the accelerating voltage (V):

$$M/e = B^2r^2 / 2V \qquad (4.5)$$

So by holding either variable B or V constant and varying the other the mass spectrum can be scanned; in practice it is usual to scan magnetically. Figure 4.12 is a diagrammatic representation. The ion source is maintained at a high vacuum to prevent unproductive collisions with air. Following a productive collision between a molecule of eluent and the electron stream from the filament, the molecular ion moves under the accelerator poten-

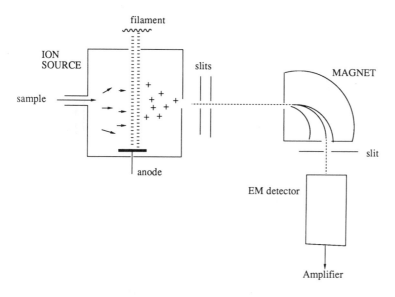

Figure 4.12   Single-focusing mass spectrometer.

tial. After focusing it passes the collector slit and the ion current is enhanced by an electron multiplier when it may be recorded.

The highly energized ions live very short lives and undergo fragmentation within the mass spectrometer, although in the great majority of examples some of the molecular ion survives. The daughter peaks which result from fragmentation are typical of the individual mother ion and can be used to identify it. Figure 4.13 shows the principal daughter peaks formed by four organohalogen compounds, the ion current of each is related to the strongest as 100%. Since chlorine has two isotopes ^{35}Cl and ^{37}Cl, in a relative abundance of 3:1 (Table 4.2) ethyl chloride has a molecular ion of mass 64 (with ^{35}Cl) and one of one-third the intensity of mass 66 (with ^{37}Cl). In this spectrum a peak is formed by loss of a methyl radical (mass 15) to leave a daughter ion of mass 49:

$$CH_3-CH_2-Cl \dotplus \;\rightarrow\; ^{\bullet}CH_3 + \;\;^{+}CH_2-Cl \qquad (4.6)$$
$$\text{lost} \qquad \text{collected}$$

Only the charged species will be collected.

In the spectrum of vinyl chloride the twin molecular ion peaks and a major peak [M-35 (37)] at 27 are seen. For trichloroethylene the record is more complex as the molecular ion has four components including a peak of low intensity (136) in which all three of the less common ^{37}Cl atoms have survived. In theory a compound with n chlorine atoms will give rise to $(n + 1)$ peaks but for polychlorocompounds like DDT and the PCBs the

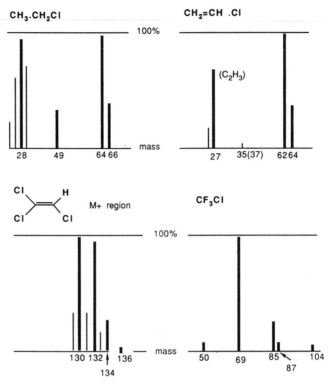

Figure 4.13 The mass spectra of some organohalogen compounds.

Table 4.2 The accurate masses of some common elements

^{12}Carbon	12.00000	^{13}Carbon	13.00335
^{1} Hydrogen	1.00782	^{16}Oxygen	15.99491
^{35}Chlorine	34.96885	^{37}Chlorine	36.96590
^{19}Fluorine	18.99840	^{14}Nitrogen	14.00307

relative ion currents of the less probable fragments may be too weak to detect.

In the spectrum of CFC 12 the molecular ion is much less stable than the most abundant CF_3^+ ion and the peak from CF_3. ^{37}Cl was not detected. Fluorine has only one nuclide of atomic mass 19 and this is why it is included in compounds used as references for accurate mass measurements (p. 75).

The link between the GLC column and the spectrometer has been made in various ingenious ways to overcome the presentation of a sample in a gas

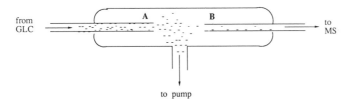

Figure 4.14   Glass jet separator.

stream to a detector operated under high vacuum. It is a common practice to divide the eluent gas stream and pass one fraction into another detector. However, only a small proportion of the remainder could be directly injected into the mass spectrometer and this would seriously impair its sensitivity. A favoured way of overcoming the incompatibility in the gas pressures is the insertion of a glass jet separator (Figure 4.14). The eluate from the GC, helium or nitrogen diffuses from the narrow orifice (A, $c.$ 0.1 mm) and is pumped from the envelope much faster than the large molecules of the analyte. Hence a highly enriched gas stream continues into the mass spectrometer via the skimmer capillary (B, $c.$ 0.2 mm).

Determination of accurate mass.   If a second electrostatic field focus is applied to the ion path in the spectrometer then much higher resolution can be obtained and fragment ion masses can be recorded at better than one-thousandth of a unit. This measurement requires a scale and this is obtained by injecting a perfluoroalkane to provide peaks of standard mass. The whole spectrum is scanned comparatively, the controlling computer then subtracts the standard spectrum and that of the analyte is printed with peaks assigned to four significant decimal places.

Given an accurate mass of say 92.03928 the possible molecular formula can be read from tables, and this corresponds to one of the isomeric chlorobutanes, C_4H_9Cl. Even allowing for experimental error this mass is clearly differentiated *at high resolution* from the nearest alternative which is glycerol, $C_3H_5O_3 = 92.04735$. Note that typical fragments are associated with each of these molecules, loss of ^{35}Cl and ^{37}Cl being unmistakable.

Another example of the value of an accurate mass would be that typical of a long chain alcohol such as nonanol, $C_9H_{19}OH = 144.15142$. This differs significantly from the accurate mass of the naphthols, $C_{10}H_8O = 144.05752$. Here the nonanol spectrum includes a daughter ion [M-18] from loss of a water molecule, while naphthol differs in that it has a major fragment ion [M-28] from loss of carbon monoxide.

The library of mass spectra includes a compendium of up to 10 major peaks of known compounds. With modern computerized data processing those obtained from the analyte can be compared with this data and the

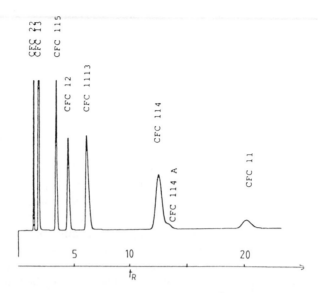

Figure 4.15 GC/MS of a mixture of chlorofluorocarbons. The t_r values in the spectrum were assigned from a parallel run using a WCOT column 25 m × 0.32 mm, temperature programmed from 30–200°C. The individual peaks were identified by a computerized comparison with a library of MS spectra (Rastogi, 1990).

best fit read out directly. An example of this in practice was the analysis of CFCs in aerosol cans (Figure 4.15). The aerosol products were examined by spraying into the collection valve of a solvent-resistant Tedler bag. Gas samples of 0.1 ml were taken from the bag for analysis and 54 out of 448 were found to contain over 1% of CFC. A proportion of these were claimed to be 'safe to the environment'.

Direct air sampling.   Mobile equipment is available (Lane, 1982) which samples air directly at a rate of 1 l/s and so gives the results in real time. For this application the ions are produced by chemical ionization rather than by electron impact. Here ions are produced by electron bombardment of air components and their charge is transferred to the substrate (M) by charge transfer, typically from hydrated protons:

$$H_3O^+ + M \rightarrow H.M^+ + H_2O \qquad (4.7)$$

This technique is well suited to the production of simple spectra in which molecular ions tend to survive because the activation is less severe. An alternative to magnetic guidance is used whereby the ions H.M$^+$ pass into a quadrupole device, with two pairs of electrodes to which one direct and one alternating potential is applied. These may be varied so that only the

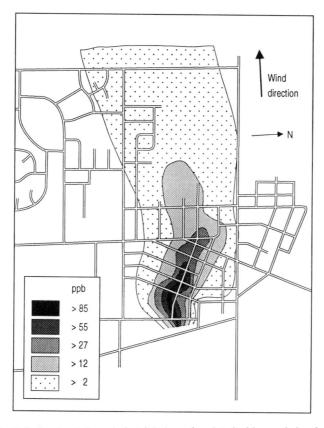

Figure 4.16   Pollution levels in an industrial plume (reprinted with permission from Lane, D.A., Environ Sci and Technol., 16, 45A, © 1982 American Chemical Society).

ion of required mass can maintain a stable path to the slit, others are lost either by collision with an electrode or by ejection from the analyser. Figure 4.16 shows isopleths, or pollution contours, collected with the equipment on a van which was driven around the neighbourhood of an industrial source. It shows an intense source of the selected ion of benzthiazole at a chimney (over 106 ppb) grading downward towards a general dispersion at 2 ppb. Clearly the application of this type of mobile analysis is a powerful way of resolving legal disputes about pollution sources.

4.4
High pressure liquid chromatography (HPLC)

Classical liquid chromatography suffers from the disadvantage that it is slow in use and fractions are eluted in substantial volumes of solvent. This is the result of imprecise loading of the analyte and the major eddy diffusion term arising from the large irregular particles of the stationary

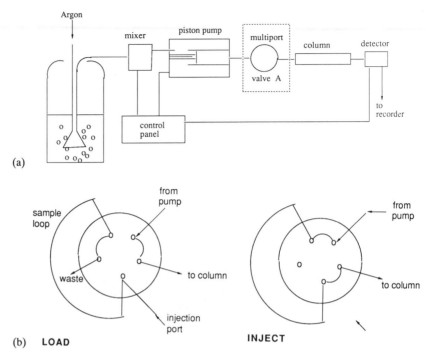

Figure 4.17 Typical components of an HPLC system (Braithwaite, 1985).

phase. It was realized early on that considerable improvement would result from efficient loading of columns of small uniform particles. However, such columns require solvent to be pumped at a stable pressure up to 6000 psi and pumps did not become available until the late 1960s. In the following years HPLC grew in popularity and capability. A block diagram of a typical system is shown in Figure 4.17.

4.4.1 The components

Columns are made of steel to withstand the high pressures and are 10–50 cm long with i.d. of 4–6 mm, although short capillary colums with an i.d. of 1–2 mm have come increasingly into use for the analysis of small quantities. *Packing material* is almost always of spherical silica of 5–10 μm diameter and pores about 10 nm. Good results can be obtained using low polarity solvents operated in the classical manner, but increasing use is made of the reverse phase mode. Here a low polarity stationary phase is attached to the column material rather in the manner of the stationary phase of GLC. To prevent loss of the stationary phase it is bonded to the silica by reaction of polar centre with a trialkylsilylchloride used in the manner described for masking in GLC (Figure 4.18).

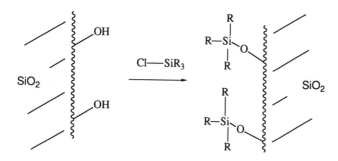

Figure 4.18 Attachment of a reverse phase to silica.

Solvents must be degassed. Figure 4.17 shows the bubbling of a pure inert gas through the solvent in the reservoir when traces of other dissolved gases diffuse into the stream and are swept out to air. Provision can be made for two or more solvents to be delivered to a mixer ahead of the column. The control can regulate the proportions and, if required, programme them by progressively enriching that of lower polarity. The plumbing must be rigorously airtight. A selection of solvents in common use is shown in Table 4.3 arranged with the weakest at the top. The viscosity is given since this bears upon back pressure and the UV cutoff because this limits UV detection.

The pump is a crucial and highly engineered component and several types are available (Ahuja, 1992). It is essential that the high pressure is delivered pulse-free in order to avoid unacceptable levels of baseline noise in the detector. For most purposes a pressure of about 1500 psi is adequate, but this will be exceeded for a high viscosity solvent.

Table 4.3 Some properties of common solvents for HPLC

Solvent	Viscosity (cp)	UV cutoff (nm)
Pentane	0.23	205
Cyclohexane	1.00	205
Toluene	0.59	285
Diethyl ether	0.24	220
Chloroform*	0.57	245
Ethyl acetate	0.45	260
Acetonitrile	0.36	210
Methanol	0.57	210

* normally contains 0.5% ethanol

Sample injection should be through a port (Figure 4.17, insert) fitted with one of a range of loops into which the charge may be injected from a syringe. This is designed so that the *full* loop may be switched into the main stream from the pump and the contents delivered in a limited volume to the column. A good system will have a minimum of free volume between the port and the column head.

4.4.2 Detectors

There have been as many as twenty types but most have specialized applications and comment is reserved here for those of widest application.

Refractive index (RI) detectors depend on dual light beams focused on two balanced photocells. One beam passes through a cell full of pure solvent while the other monitors the flow from the column; once eluant enters the latter the RI changes and the photocells go out of balance. This type of detector can be used widely but its sensitivity is limited to about 300 ng/ml.

Ultraviolet detectors are of great importance although their use is restricted by solvent cutoff and confined to molecules which absorb UV above this point. In practice this limitation is not so serious because many polluting substances fall within this category. It is also often possible to form a derivative of the substrate which is UV absorptive. For example, aliphatic alcohols react readily with aromatic acid chlorides.

$$R\text{-OH} + \quad \underset{Cl}{\overset{O}{\|}}C\text{-}\langle\bigcirc\rangle\text{-NO}_2 \longrightarrow \underset{RO}{\overset{O}{\|}}C\text{-}\langle\bigcirc\rangle\text{-NO}_2 + [HCl] \qquad (4.8)$$

The sensitivity towards a solute molecule can be gauged from the molecular extinction coefficient (ϵ) at a maximum. The detector can be tuned to respond to this narrow wavelength band and when absorption occurs in the eluent path the balance with a pure solvent passing is lost, in the manner of the RI detector.

The relation between incident (I_0) and emergent light is given by:

$$\log I_0/I = \epsilon\, l.c \qquad (4.9)$$

where l = length of light path (cm) and c = molar conc./l. From this we see that ϵ is equivalent to the reciprocal of the concentration required to reduce the incident light to one-tenth. For a nitrobenzene standard ϵ = 9000 at 251 nm and if we assume that a 50 ng sample enters the detector with a 1 cm path in a volume of 0.2 ml the change in through light is given by:

$$\log I_0/I = 9 \times 10^3 \times 50 \times 10^{-9} / M \times 0.2 \times 10^{-3}$$

Table 4.4 Some polycyclic aromatic hydrocarbons

Benzo-[a]-pyrene	Strongly carcinogenic and mutagenic at 5 μg/plate
	Occurs at 2 μg/100 cigarettes
	Fuel oil at 10 mg/t
	London air 20 ng/m^3
Chrysene	Weak carcinogen
	Occurs at 6 μg/100 cigarettes
	Indoor air 4 ng/m^3

with M = 123; log I_0/I = 0.018 whence I_0/I = 1.04 which would cause a detectable 4% change.

The structures of a group of polycyclic aromatic hydrocarbons are shown in Figure 6.4. They are original products of coal tar distillation and are present in diesel exhaust and tobacco smoke. They are the subject of concern because the group includes cancer-inducing substances. The activity of benzo-[a]-pyrene, the most toxic, is compared with that of the weakly active chrysene in Table 4.4. These compounds absorb UV light more intensely than nitrobenzene because of their extended conjugated systems and the UV detector is very sensitive to them (Figure 4.19). These compounds and some others can be detected with still greater sensitivity by inducing their fluorescence with a laser light source.

Electrochemical detectors are extremely sensitive and respond to as little as 1 ng/ml. They depend for their action on the electro-oxidation or reduction of the eluent and the potential change which then occurs in the cell. Unless they are derivatized, ethers, alcohols and carboxylic acids are not sensed.

The criteria for efficiency are similar to those discussed for GLC, but it should be noted that eddy diffusion is limited by particle size and longitudinal diffusion is less important because it occurs so much more slowly in the flowing liquid phase than in the gas.

Speed is of the essence and consequent inefficiencies in mass transfer are more than compensated for by the wide range of solvents available, especially as they can be mixed and varied continuously throughout the elution. These advantages mean that separations can be achieved on a short column (*c.* 25 cm) that would require many metres of a GLC column. Comparisons are, however, odious where GLC and HPLC are concerned since both are essential for the analysis of the wide range of environmental samples, but some pros and cons are given in Table 4.5.

4.4.3 Analysis of polluted air

Phenols were trapped by bubbling air through a weakly alkaline solution containing a diazonium salt (**9**) and the solution of diazotized phenol

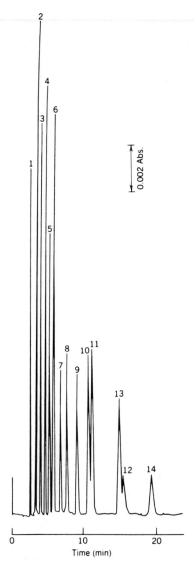

Figure 4.19 Separation of PAH on a microcolumn of silica ODS. Column 10 cm ×
0.25 mm, flow rate 3 μl/min, detection at 254 nm. Sample: 1, benzene; 2, naphthalene; 3,
biphenyl; 4, fluorene; 5, phenanthrene; 6, anthracene; 7, fluoranthene; 8, pyrene; 9, *p*-
terphenyl; 10, chrysene; 11, 9-phenylanthracene; 12, perylene; 13, 1,3,5-triphenylbenzene;
14, benzo-[a]-pyrene. Wavelength of UV detection; 254 nm. Injection volume: 0.02 μl
(Takeuchi, 1981; Ahuja, 1992).

(equation 4.10) was made up to a standard volume and an aliquot injected
into the HPLC. Figure 4.20 shows the separation of those phenols which
occur in industrial emissions, car exhausts and tobacco smoke.

Table 4.5   Comparative behaviour

Feature	GLC	HPLC
Speed	Adequate	Superior
Involatiles	Fails	Acceptable
Metal ions	Inapplicable	Acceptable
Gases	Acceptable	Inapplicable
Temperature programming	Acceptable	Unnecessary
Flowing phase	Limited, inert	Extensive, active
Detection	Excellent	Depends on substance

$$Ph\text{-}OH \;+\; {}_2ON\text{-}\!\!\langle\rangle\!\!\text{-}N\!\equiv\!N^+\,BF_4^- \;\longrightarrow\; {}_2ON\text{-}\!\!\langle\rangle\!\!\text{-}N\!=\!N\text{-}\!\langle\rangle\!\text{-}NO_2$$

9

(4.10)

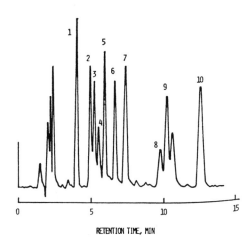

Figure 4.20   Separation of phenolic pollutants. 20 cm column in reverse phase, 85%
MeOH:15% water, at 1.1 ml/min, 1500 psi, detection at 365 nm. Phenols 50–400 ng. 1,
Phenol; 2, *m*-cresol; 3, *o*-cresol; 4, α-naphthol; 5, 3,5-xylenol; 6, 2,3-, 2,5- and
2,6-xylenol; 7, *p*-cresol; 8, β-naphthol; 9, 3,5-xylenol; 10, 2,4-xylenol (reprinted with
permission from Kuwata, K. *et al.*, *Anal. Chem.*, **52**, © 1980 American Chemical Society).

Amines in ambient air can be detected at concentrations as low as 0.1 μg/m^3
although volumes of the order of 100 l must be drawn through an
adsorption column impregnated with phosphoric acid. Once the sample
has been collected the free amines are released by addition of alkali and
then derivatized to ensure UV detection. One suitable method is conversion into the *m*-toluamides (**11**) by reaction with *m*-toluoyl chloride (**10**) as
shown in equation 4.11. Figure 4.21 shows typical responses of reference

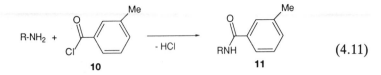

$$(4.11)$$

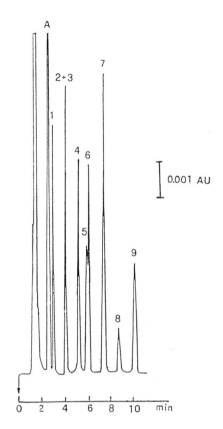

Figure 4.21 Separation of aliphatic amine air pollutants. A = ammonia (excess); 1, methylamine (55 pmol); 2 and 3, ethylamine (61 pmol) + dimethylamine (40 pmol); 4, allylamine (53 pmol); 5, isopropylamine (47 pmol); 6, n-propylamine (48 pmol); 7, ethylenediamine (58 pmol); 8, diethylamine (38 pmol); 9, n-butylamine (40 pmol) (Possanzini, 1990).

samples of those amines found in polluted air. The sensitivity may be gauged from peak 6 which corresponds to 48 picomol of *n*-propylamine and peak 9 due to *n*-butylamine (40 picomol). Isocratic elution, that is with a solvent of constant composition, is quicker than gradient elution but here fails to separate ethylamine from dimethylamine (peaks 2 and 3).

This technique is capable of detecting pollutants from tobacco smoke. A typical value for methylamine is 1.6 μg/m^3.

4.4.4   Analysis of polluted water

HPLC columns can be used for trace enrichment of these samples. A measured volume of water is drawn through the column when the organic contaminants are taken up preferentially by the ODS layer. Once sufficient volume has been passed through the column a graded change of solvent is made leading to elution of the analyte. In practice it is best to concentrate the sample in a short pre-column which is then connected to an analytical column by replacing one loop of the injector (Figure 4.17). To concentrate the charge the pre-column is reversed so that the charge goes directly into the analytical column. Some diffusion of the sample is inevitable in this procedure and as can be seen (Figure 4.22) broad peaks are obtained. Note that the peak at 1.37 min marks the point in time when the analyte solvent enters the detector cell. It can be seen from their unsaturated structure (Figures 6.11, 6.13) that all five of the component pesticides absorb UV light in the region of 220 nm, but the maxima will not coincide at this wavelength (Table 4.6) and the molecular extinctions will differ,

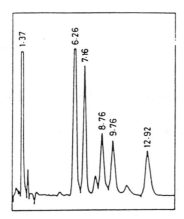

Figure 4.22   Separation of a pesticide mixture after trace enrichment. 5 cm pre-column of 10 μm ODS silica eluted with MEOH/water (3:1) through a 25 cm ODS analytical column at 2 ml/min with UV detection at 220 cm (Braithwaite, 1990).

Table 4.6   Individual pesticides from trace enrichment

Compound	Peak (t_r)	UV max. (nm)	Content (μg)
Methoxychlor	one, 6.26	—	13
Dieldrin	two, 7.16	216	23
Heptachlor	three, 8.76	213	22
DDT	four, 9.76	206	14
Aldrin	five, 12.92	213	17

hence for quantitation each response must be compared with that of a reference standard.

4.4.5   Trace enrichment followed by GLC analysis

An example of this application is the detection of chlorophenols accidentally released from a sawmill into lake water. A sample (100 ml) taken downstream was injected through a plug of ODS silica, which was then reversed and eluted with acetone. Use of this powerful solvent ensures that most of the analyte is eluted in a volume of only 4 ml and since UV is not used for detection there will be no interference.

To ensure good resolution the polar OH group in the phenols (e.g. **12**) was acetylated with acetic anhydride. Excess of this reagent was removed

$$\text{⬡—OH} \quad + \quad (\text{Ac})_2\text{O} \quad \longrightarrow \quad \text{⬡—OAc} \quad + \quad \text{AcOH} \qquad (4.12)$$

12

by hydrolysis and the *o*-acetates extracted into 1.0 ml of hexane. This procedure gives good resolution (Figure 4.23) since the analyte is contained in a small volume.

Quantitative results were obtained by the addition of the same amount of 2,6-dibromophenol to all samples when it acts as an internal standard (i.s.). Then if a substance (*s*) gives rise to a peak of area (A_m) and the standard to a peak of area (A_{im}) then the concentration (C_m) of the substance is given by:

$$C_m = A_m/A_{im} \times C_{im} \times R_{im}/R_m \qquad (4.13)$$

the ratio of the response factors R_{im}/R_m of the internal standard and each component of the analyte must be determined independently by plotting the relative peak areas for a series of mixtures of known composition (also see p. 66). The result of the analysis is given in Figure 4.24.

4.5
Pollution by metals – atomic absorption spectroscopy (Lajunen, 1992)

Attention will be concentrated on this principal method of determining the levels of heavy metals in environmental samples.

4.5.1   Historical

The technique had its beginnings in 1814 with Fraunhofer's observation of line spectra in the sun. In 1859 Kirchoff and Bunsen established the key

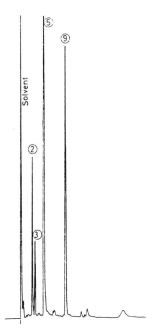

Figure 4.23   GLC of acetyl chlorophenols from polluted lake water. A 25 m × 0.2 mm SE–30 quartz capillary column was used at 200°C with a helium flow rate of 0.6 ml/min. In order to avoid overloading, the injection was split so that only 1/20 entered the column, the rest going to waste (Renberg, 1981).

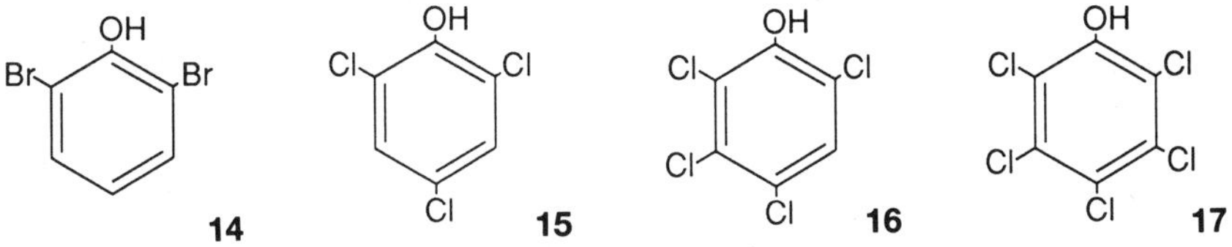

Compound	Peak	Concentration
14 2,6 - dibromophenol	3	0.054 µg / ml
15 2,4,6 - trichlorophenol	2	0.50 µg / l
16 2,4,5,6 - tetrachlorophenol	5	1.8 µg / l
17 pentachlorophenol	9	0.3 µg / l

Figure 4.24   Structure and concentrations of chlorophenols in lake water. Peak 2: 2,4,6-trichlorophenol (0.12 µg/ml); peak 3: 2,6-dibromophenol (internal standard, 0.54 µg/ml); peak 5: 2,4,5,6-tetrachlorophenol (0.11 µg/ml); peak 9: pentachlorophenol (0.14 µg/ml).

principle that 'Matter absorbs light at the same wavelength at which it emits light'.

Atomic absorption (AA) was first used by Woodson in 1939 for the analysis of mercury in air. In 1955 Walsh showed how metal ions could be reduced to the parent atoms within flames and that their concentration could then be found by the absorption of monochromatic incident light, although it was almost 10 years before a commercial instrument became available. Since then the use of AA spectrometers has expanded rapidly and with the introduction in 1970 of the plasma source there has been an increase in the related method of atomic emission spectroscopy (AES).

4.5.2 Basic theory of atomic absorption and emission

When a solution containing metal ions is introduced as fine droplets into the gas stream entering the burner (Figure 4.27), the metal ions are reduced to neutral atoms by electron capture.

Figure 4.25 is a simplified diagram of the energy levels in the lithium atom which has a pair of electrons in the 1s orbital of lowest energy and a single bonding electron in the higher 2s orbital. When the atom is excited

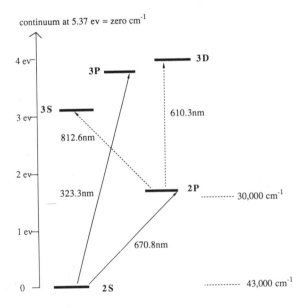

Figure 4.25 Energy levels for the lithium atom. 1 unit on voltage scale = 8000 cm^{-1}. A low energy transition corresponds to a long wavelength. For the 2s → 2p transition, wavenumber = 1/wavelength (cm^{-1}) = 10^7/670.8 = 14 900 cm^{-1}.

by an input of energy, as experienced in the flame, the electron in the $2s$ orbital may enter an orbital of higher energy. The various levels are restricted by quantum theory and subject to Planck's equation (4.14):

$$\text{Energy} = h \,/\, \text{line wavelength} \qquad (4.14)$$

where h is Planck's constant. The energy difference between the $2s$ and the $2p$ orbitals is almost 2 electron volts (eV) and this precise quantum will be absorbed from the flame to promote an electron to the upper level. We see from Figure 4.25 that this corresponds to a line in the spectrum of wavelength 670.8 nm. The lines for three other transitions are shown corresponding to wavelengths of 323 ($2s{\rightarrow}3p$), 812 ($2p{\rightarrow}3s$) and 610 nm ($2p{\rightarrow}3d$) so that the lower energy corresponds to the longer wavelength.

The frequency scale on the right of Figure 4.26 is in reciprocal centimetres and bears an inverse relation to the electron volt scale on the left. For the $2s{\rightarrow}2p$ transition:

frequency $(\text{cm}^{-1}) = 1$
wavelength $(\text{cm}) = 10^7$
$670.8 = 1.49.10^4 \text{ cm}^{-1}$

The strongest line in the lithium spectrum is that arising from promotion from the lowest $2s$ level to the next highest $2p$ level. The intense red colour of the flame arises from emission at 670 nm when the excited electron falls back into the lower level. It follows from Kirchoff and Bunsen's principle that we may evaluate the concentration of atoms in flames either by the intensity of absorption or emission at a chosen wavelength.

In practice only a small proportion of atoms are excited at any one time, even for the most probable transitions, and Table 4.7 shows how this rises with temperature for the easily excited golden yellow line of sodium (589 nm) and for the blue calcium line (423 nm). It follows that flame temperatures of at least 2000 K are needed if sufficient sensitivity is to be achieved and this is possible when acetylene burns in air. Acetylene/oxygen flames exceed 3000 K and acetylene/nitrous oxide reaches 3400 K. At high temperatures an electron may be expelled altogether and this is a source of

Table 4.7 Variation of excited to ground state atoms with temperature

	2000 K	3000 K	4000 K	5000 K
Sodium (589 nm)	1 in 10^5	6 in 10^4	4 in 10^3	2 in 10^2
Calcium (423 nm)	1 in 10^7	4 in 10^5	6 in 10^4	3 in 10^3

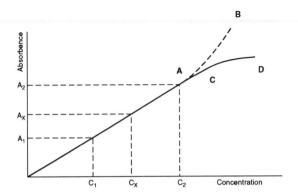

Figure 4.26   Typical calibration curve for AAS.

inefficiency as the ions are not recorded. An input of 5.37 eV is required to ionize lithium but potassium with an ionization potential of 4.34 eV gives ions more readily.

4.5.3   The Lambert–Beer Law

This states that at a given wavelength the absorbance is proportional to the number of absorbing species in the light path. With absorbance defined as $\log I_0/I$ where I_0 is the energy emitted and I is the energy entering the detector, we have:

$$\log I_0/I = a.\ l.\ c \qquad (4.15)$$

Where a = a constant, l = length of light path and c = the required concentration. This relation is given in another form in equation 4.9. Therefore, by plotting absorbance against a series of samples of known concentration a calibration curve can be produced (Figure 4.26) from which any atomic concentration can be read, given a known detector response.

At the normally low excitation levels the law is usually obeyed and a linear plot results. However, at high concentrations some light may pass through unabsorbed and the plot then turns along CD towards the concentration axis. If ionization is occurring, the plot turns along AB as ionization is reduced at higher concentrations. Provided it has been verified that the curve is linear at lower concentrations, an unknown concentration (C_x) can be obtained from the its absorbance (A_x) by comparison with measurements for two reference standards (C_1, A_1) and (C_2, A_2).

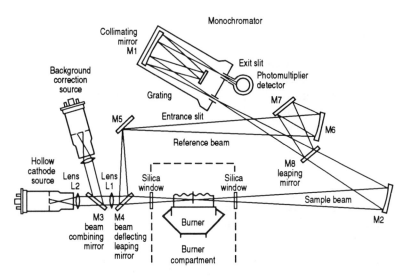

Figure 4.27   Atomic absorption spectroscopy – diagram of the basic instrument
(UNICAM Ltd).

4.5.4   Instrumental details (Figure 4.27) (Unicam)

The lamp operates in the range of 100–400 V and is filled with either argon
or neon. When high energy electrons leave the cathode they generate
positive ions of the inert gas (Ar^+ or Ne^+) which are accelerated to strike
the *hollow* cathode and sputter atoms from it. These atoms are excited in
their turn as was shown for lithium (Figure 4.25) and as these fall back into
the ground state the emission spectrum is produced and escapes through
the hollow cathode to produce the incident beam. A range of these lamps
(HCL) is available and the one whose cathode element corresponds with
the one to be determined is selected.

The monochromator for AAS need not normally be capable of high
resolution because the emission of the lamp matches the spectrum of the
analyte. It may be adjusted to select any suitable spectral line and should
be able to select emitted lines separated by 0.2 nm (but see section 4.5.5).

The detector is a photomultiplier tube which responds to the incident
emission by releasing electrons from its cathode. This initial current is
enhanced by interaction with a series of anodes and finally used to activate
a recorder.

The recorder may be a simple ammeter but modern instruments can
store and reveal readings on a video display.

Double beam operation is used (Figure 4.27) whereby the output from
the lamp is divided by a rotating mirror-chopper combination M_3, M_4 into
a sample beam M_4, M_2 which passes through the flame and a reference
beam M_5, M_6 which is recombined after passage. The systems electronics

takes the ratio of the two I_{ref}/I_{sample} so balancing out any transitory changes in the source.

The system avoids possible error with a single beam instrument, where the source output may change between measurement of the standard and measurement of the sample. Double beam operation has the disadvantage of reducing the incident light on the sample to half and it cannot compensate for variations within the flame. For emission measurement the lamps are turned off as the flame is the energy source.

4.5.5   Interferences

Apart from variations in the source, other factors may introduce errors of measurement and these are summarized in this section.

Stable compounds may be formed by oxidation in the flame to refractory oxides which are incompletely atomized. Metals which behave in this way include Al, Ca, Mg and Ti; phosphates and silicates of alkaline earth metals may also lead to under-estimation. The remedy is to employ the hotter N_2O/acetylene flame.

Ionization will increase in a hotter flame leading to a reduction in the concentration of unionized absorbing atoms of the analyte. It may be countered by the addition of a metal which is more readily ionized, for example the ionization of lithium is suppressed by the addition of 0.1% of potassium chloride (cf. p. 90). The accurate determination of lithium is required in medical research because this metal is a powerful depressant of the central nervous system.

Spectral interference is rare for AAS given the well resolved output from the HCl so that coincidence in the wavelength of absorbance of two species is unlikely. Should this occur, as for that between the lines for Cu at 213.859 and Zn at 213.856 nm, then Cu can be determined at 217.894 nm. This type of interference is significant for plasma sources (p. 96) because the high temperatures intensify many lines that are weak in the temperature range 2–3000 K.

Non-specific absorption arises from absorption and scatter of incident light by species other than the analyte and can be allowed for by making a background correction. The Smith–Hiefte method is one way of doing this; a measurement is taken at the normal lamp operating current and then another well above this current. At the higher level the emission lines are broadened so that most of the incident light is not absorbed by the analyte. Only the non-specific absorption is recorded and it may then be subtracted from the normal measurement.

Physical interference may arise if the standard and sample are taken up at different rates into the flame from the nebulizer. This can be avoided by preparing standard and sample in a common solvent.

Matrix interferences are related to compound formation between the analyte and other species in the sample. In the following example this interference is overcome by equating the matrix concentration in both standard and sample solutions.

4.5.6   The determination of sodium in concrete by AAS (Adams, 1992)

One cause of failure in concrete results from the uptake of sodium chloride used for de-icing and its conversion into the strongly alkaline oxide Na_2O. One example of this problem is to be found in raised sections of the M6 motorway around Birmingham in the UK.

The analysis is complicated by matrix interference because the critical oxides are at a low concentration (Na_2O normally 0.1%, K_2 normally 0.5% with CaO about 30%). The large excess of calcium interferes with the determination of both alkali metals and its effect is counteracted by the addition of aluminium as nitrate. As a result calcium becomes combined as the aluminate, $CaAl_2O_4$; this is stable at the temperature of the air/acetylene flame and so the calcium is not atomized and measurement of Na and K may proceed. This is an example of an advantageous chemical interaction.

The experimental procedure entails the extraction of powdered concrete or a sample of cement (c. 0.5 g) with boiling dilute nitric acid. The cool extract is diluted to 500 ml in a volumetric flask and a pipetted aliquot (25 ml) mixed with aluminium nitrate (10 ml $\equiv$ 0.2 g Al) and made up to 100 ml. These dilutions are chosen so as to give concentrations of analyte on the linear section of the standard plot, assuming the low levels given above.

Aluminium salts always contain some sodium and allowance for this must be made by preparing a blank solution of the aluminium nitrate, using the same dilution procedure, and obtaining an absorbance reading (B_{Al}) for its sodium content.

Because of the matrix interference a standard blank must also be prepared having the same concentrations of Al and Ca as the sample. It will give a further absorbance figure for the standard blank B_{stand}.

For a sample (0.5009 g) of ordinary portland cement the following absorbances were obtained (cf. Figure 4.26):

	Absorbance
Portland cement	0.420
Aluminium blank (B_{Al})	0.026
By difference the sample absorbance	0.394
Na_2O standard 0.5 µg/ml absorbance	0.272
Standard blank (B_{stand})	0.090
By difference standard absorbance	0.182

Therefore, by proportion, Na in cement $0.5 \times \dfrac{0.394}{0.182}$ µg/ml and the Na content in the original extract of 500 ml from 0.5009×10^6 µg $=$
$$\frac{0.5 \times 0.394 \times 500 \times 100}{0.182 \times 0.5009 \times 10^6}\% = \mathbf{0.109\%}$$
A figure which falls within acceptable limits.

4.5.7   Sample preparation

In many analyses the method used for cement will be satisfactory and if need be concentrated nitric acid can be used to break down organic material and to oxidize metals which resist other acids. For highly resistant samples such as plant seeds a mixture of $HNO_3:HClO_4:H_2SO_4$ as 3:1:1 by volume is used, the sulphuric acid is added to ensure that on concentration the lower boiling perchloric acid evaporates first because of the risk of explosion should it accumulate.

Solutions can also be obtained by fusion of material with fluxes such as sodium carbonate or sodium peroxide. This method suffers from the disadvantage of introducing a large concentration of ions into the matrix.

4.5.8   Precision and accuracy of measurement

A distinction must be made between these two factors.

Precision refers to the range within which results may be reproduced, which for AAS should be 0.5%, provided that the measurements are made at concentrations well above the detection limit.

Accuracy refers to the difference between the true and measured result. It will be affected by random errors arising during sampling, preparation, dilution and injection of the sample.

Measurements could be made with high precision but would still be inaccurate if a systematic error was being made, for example failure to recognize spectral interference.

Analysis of small samples.   Flame AAS aspirates the sample at about 4 ml/min and this is acceptable when a large volume is available, as in the example above. However, consider the determination of cadmium, a rare metal which can cause kidney damage if taken in excess of WHO maximum tolerable levels of 400 µg/week. Although this metal occurs at average levels of only 0.015 µg/g in UK soils it is concentrated by plants, especially vegetables such as lettuce and spinach, reaching the order of 1 µg/g in leaves of dry weight some 50 mg; such a sample therefore contains $1 \times 50/1000$ µg or 1/20 µg.

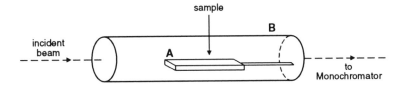

Figure 4.28   Simplified diagram of a graphite furnace.

The detection limit for Cd by flame AAS (Figure 4.30) under ideal conditions is <1 µg/l and the leaf sample contains 1/20 of this, enough to provide 50 ml at the detectable concentration. To ensure precision, at least ten times this concentration is necessary, that is a sample volume of only 5 ml, which is inadequate. This type of problem has been overcome by the use of a graphite furnace to atomize the sample.

4.5.9   Graphite furnace AAS

Figure 4.28 is a simplified diagram. It was shown by L'vov that atomization could be improved by aspirating the sample in a stream of inert gas on to the graphite platform (A), within the heated graphite cylinder (B). The platform is supported in poor thermal contact with the outer cylinder; this ensures that the platform temperature lags behind, so that when the analyte is atomized it enters a stable gas phase at a steady temperature.

A temperature programme is followed which is repeated for each sample. A typical sequence would be:

1. Slow drying up to 400 K for 30 seconds (matrix modifiers may be added here if needed).
2. Thermal pretreatment to bring the temperature to about 700 K. This allows for the dispersal of volatile decomposition products which would interfere if present during the absorption measurement.
3. The gas flow is stopped so that the atomized analyte is not swept out of the light path.
   Cd is volatile and its spectrum will appear in only 0.3 s; the absorption will be measured by summation over the next 2–3 s.
4. The furnace temperature is raised 100°C above the highest atomization temperature to clean the system before cooling.

A valuable application of the furnace is in the atomization of small solid samples.

4.6
A plasma source

The plasma is an electrically excited volume of gas which includes a significant fraction of ionized atoms or molecules. Figure 4.29 shows the

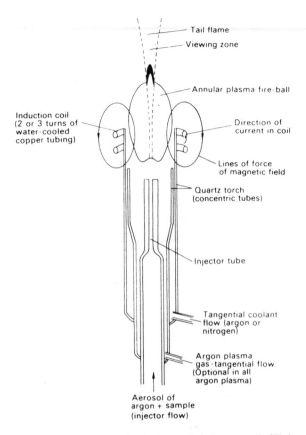

Induction coil
(2 or 3 turns of
water-cooled
copper tubing)

Tail flame

Viewing zone

Annular plasma fire-ball

Direction of
current in coil

Lines of force
of magnetic field

Quartz torch
(concentric tubes)

Injector tube

Tangential coolant
flow (argon or
nitrogen)

Argon plasma
gas-tangential flow
(Optional in all
argon plasma)

Aerosol of
argon + sample
(injector flow)

Figure 4.29   Diagram of the inductively coupled plasma torch (Ebdon, 1982).

most common form with inductively coupled plasma (ICP). This was researched by Greenfield and Fassel and consists of three concentric tubes. The outer is a cooling tube, the centre supplies argon to feed the plasma and the inner injects argon with analyte as an aerosol. The exit is surrounded by an RF induction coil which induces an oscillating magnetic field through the coil. When the torch is seeded by a spark electrons are released and, given sufficient power, argon ions are formed by collision (cf. the mass spectrometer, p. 71) to form the plasma. This is rich in ions at a temperature of 10 000 K, although the viewing zone above the coil is at 6000 K. These conditions are not suitable for AAS and hence emission is determined (ICP–AES).

AES has the advantage that the intensity is linear and is not attenuated as in AAS and the ICP ensures that even refractory elements such as uranium and boron are atomized.

Boron is an essential plant nutrient; cereal crops require soil with between 5–20 μg/g. The range in the UK is from 7–70 μg/g but toxic symptoms appear at levels only slightly above this. In semi-arid areas irrigation water can carry toxic concentrations of boron which arise from contact with the water-soluble minerals kernite ($Na_2O. 2B_2O_3. 4H_2O$) and colemanite ($2CaO. 3B_2O_3. 5H_2O$). In these regions the sunflower crop is at risk; below 35 μg/g the plant suffers from deficiency but an upper level of 150 μg/g should not be exceeded. The availability of ICP–AES is essential for such studies.

Spectral interference is significant in ICP–AES because many more lines are excited and high resolution is required in the monochromator. Instruments can be fitted with a polychromator having numerous exit slits, which can be adjusted to select a range of elemental emissions simultaneously. This equipment is expensive to install and operate and is normally found where routine multi-element analysis is required. For the general user AAS remains the technique of choice especially as it is comparable in sensitivity and even superior: compare Mg, Ag and Pb on the logarithmic scale of Figure 4.30.

4.6.1 ICP-mass spectrometry

The production of positive ions in the plasma allows their identification and measurement in ways that have been mentioned (p. 71). There can be no spectral interference but interfering ions may be produced in the plasma, for example if nitric acid is aspirated $(argon.N)^+$ ions will be formed having the same mass as ^{54}Fe and sulphuric acid generates SO_2^+ coincident with ^{64}Zn.

The operation of this instrument requires considerable expertise but it can detect elements at very low limits (Figure 4.30).

**4.7
Analytical quality
assurance**

Quality assurance is particularly important in environmental monitoring where wide variations in concentrations of pollutants are expected in large batches of samples. Analysis of replicates, with at least 20% of samples duplicated, is essential as well as the use of in-house standards and certified reference materials (CRMs). Several organizations, including the European Community Bureau of Reference (BCR), International Atomic Energy Agency, US Bureau of Standards, US Geological Survey, CAN-MET in Canada, and others offer certified materials including various types of soils, some amended with sewage sludge, plant and animal tissues, food materials and others. Appropriate CRMs should be selected on the basis of being as close to the monitoring samples as possible in both matrix and concentrations of pollutants. Owing to their cost it is often not practicable to use aliquots of these in every batch of samples and so most

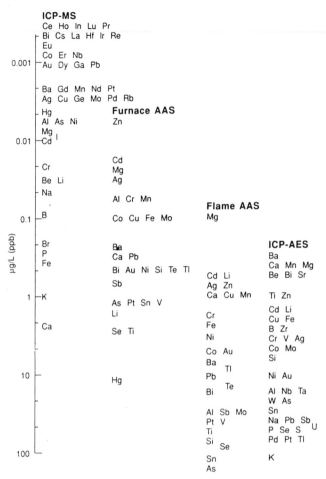

Figure 4.30 Comparison of detection limits (Slavin, 1992).

laboratories involved in the analysis of large batches use in-house reference materials comprising large bulk samples (tens of kilograms in the case of soils) of thoroughly homogenized material that has been repeatedly analysed (often by other laboratories as well) and calibrated against a relevant CRM. (A range of soil CRMs for heavy metal analysis is listed in Alloway (1990).)

The results for duplicates and CRMs or in-house reference materials analysed under the same conditions as the monitoring samples should be quoted with the results of the environmental samples. Under normal circumstances, it is expected that the values for duplicates should be better than ± 10% and the means for CRM replicates should fall with ± 10% of certified values for the sample. The use of CRMs provides an indication of quality control at all stages through the analysis.

4.8.1   Introduction

Having discussed some of the methods used for the analysis of organic and inorganic environmental pollutants, it is necessary to consider briefly the rationale for collecting the samples to be analysed. It should be remembered that no matter how accurate and precise the analyses, if the samples are not appropriate the results will be meaningless. Therefore adequate time should be devoted to preliminary planning and investigation of the areas to be monitored in order to ensure the most cost effective sampling regime. The old adage 'what your purpose is determines the method you choose' is particularly pertinent to environmental monitoring because there are many different ways of monitoring.

In general, monitoring is carried out to determine (adapted from Holdgate (1979) and Hewitt and Allott (1992)):

1. The substances entering the environment, the quantities involved, the sources from which they came, and their spacial distribution.
2. The effects of these substances on humans, crops, livestock, ecosystems and structures.
3. Any trends in the concentration of these substances over time and the reasons for them.
4. To what extent the inputs, concentrations, effects and trends of pollutants can be modified, by what means and at what cost.
5. To establish baseline concentrations of possible pollutants for comparison with data for other locations and data for the same location after the commencement of polluting activities.
6. To assess the need for legislative controls on emissions of pollutants and ensure compliance with existing regulations.
7. To activate emergency procedures in areas of high risk (e.g. accidental emissions from chemical plants or the accumulation of hazardous levels of pollutants from road traffic).
8. To determine the suitability of water and land resources for various proposed uses.

On the basis of the objectives, a series of decisions needs to be taken about the essential components of a monitoring programme as shown in the flow chart by Hewitt and Allot (1992) (Figure 4.31).

Monitoring can be concerned with:

1. sources – investigating the nature and rates of emissions into air or water, or fall-out onto land;
2. transport mechanisms – concentrations of pollutants, distance transported and whether they have been chemically transformed;
3. targets – the amounts and speciation of pollutants reaching or entering individual targets (humans, animals, plants, ecosystems or structures).

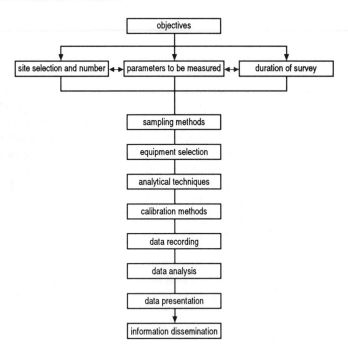

Figure 4.31   Flow chart of the essential components of a monitoring system (reproduced from Hewitt and Allot (1992)).

4.8.2   Monitoring emissions

Emissions of pollutants from a source can be of three types: (i) *planned emissions* which normally leave either a chimney or a pipe in concentrations which fall within a range that is either licensed, or at least planned for, (ii) *fugitive emissions*, which are not planned but happen as a result of leakages of fumes or liquids from within processes but nonetheless can sometimes constitute significant inputs of the pollutant into the environment, and (iii) *accidental emissions* which can range from minor accidental releases due to accidentally opening valves, to explosions which release vast amounts of pollutants into the environment, as in the accidents at chemical plants in Séveso (see section 6.5) and Bhopal (atmospheric pollution), or the fire at the Sandoz chemical stores in Basle (river pollution).

When planning to monitor emissions or discharges from known or suspected sources, the range of pollutants involved, their rates of emission and fluctuations over time are of crucial importance. Marked variations can occur in some industrial plants where pollutant-emitting processes are only carried out intermittently, or at certain times. In the case of air and water whose composition will fluctuate markedly in response to variations in emissions, it is important to ensure that the monitoring system will run for sufficient time to detect these variations. On the other hand, soil and, to a certain extent, river sediment will provide an indication of the overall extent of pollution as reflected in fall-out on the soil or sorption in the sediment because these media sorb most pollutants. In all cases, it is essential that an adequate number of sampling sites are used in order to provide data for both the polluted media and the background levels at sites unaffected by the pollution being monitored. With air sampling, variations in wind direction need to be allowed for and sampling sites should be along transects in North, South, East and West directions where practicable. Where sources are sited on coastlines and other geographically limiting locations this may not always be possible. The data obtained during a monitoring survey of atmospheric emissions will provide comparative values for polluted and non-polluted air because not all samples will be in the path of the pollutants at any one time. Meteorological data for the area would be useful because they provide information on the direction of prevailing winds and the range of weather conditions normally experienced at the site.

Atmospheric sampling can include total deposit gauges which collect both rainfall and dry deposition, samplers which pump a known volume of air through selected types of filter (including various mesh sizes for particulates and tubes packed with adsorbents to sorb VOCs), Dreschel bottles containing various absorbents for bubbling air through to absorb gases, and automated apparatus carrying out continuous sampling and analysis for chemicals such as O_3 and NO_2. Passive air samplers comprising plastic tubes filled with selective gels to absorb pollutants can also be used for some pollutants including NO_2 and hydrocarbons. These are much less precise than pumped air samplers/analysers but have the advantage of being inexpensive and thus allow much more intensive coverage. However, the tubes have to be left exposed to the air for about a week and so can only be used to indicate average concentrations of pollutants. Therefore, they are most useful when used in integrated surveys together with pumped air filters. The most commonly used methods for the analysis of many pollutants in air are given in Table 4.8.

Surface water sampling to monitor releases of pollutants from a source should include an adequate number of sampling points above the outfall pipe to provide reliable control data. Sampling locations below the outfall should allow for both the mixing of the pollutants with the river water and also the effects of tributary drainage downstream. If this tributary is heavily polluted it could lead to confusion about releases from the source under investigation. It is important to maintain a constant depth of sampling for surface waters and specialist equipment is available to collect grab samples of water at preselected depths. A wide range of analytical methods is used for water pollutants and many industrial plants and river

Table 4.8 Some methods used for the analysis of air pollutants (adapted from Harrison (1990))

Pollutant	Analytical method
SO_2	Absorption in H_2O_2 and titration
	Absorption in tetrachloro-mercurate and spectrophotometry
	Flame photometric analysis
	Gas phase fluorescence
NO_x	Chemoluminescence with ozone
Total hydrocarbons	Flame ionization analysis
Specific hydrocarbons	Gas chromatography/flame ionization detector
CO	Catalytic methanation – flame ionization detector
	– electrochemical cell
	– non-dispersive infrared
Ozone	Chemiluminescent reaction with ethene
	UV absorption
PAN	Gas chromatography – electron capture detection
Particulates	High volume sampler – filters extracted or digested analysed by AAS and GFAAS
	Reduction in reflectance of glass slide

sampling locations have automated continuous analysis of selected pollutants. Metals in solution and suspended particulates are normally determined by AAS, usually after a solvent extraction pre-concentration process, but GFAS can also be used. Organic pollutants are analysed by the methods listed earlier.

The monitoring of soil is more straightforward than that of air and water because the concentrations of pollutants in soils are less subject to temporal variations. This means that the timing of sampling is less critical, but it is still important to prevent contamination of samples in the same way that it is for air and water samples. However, soils are inherently heterogeneous with regard to many properties and pollutants. In the case of metals, the rock forming minerals in the soil parent material vary considerably in metal composition and so even relatively unpolluted soils differ spacially in metal composition (see Table 5.13). However, most soils in the world have been polluted at least to a slight extent, from atmospherically deposited pollutants, fertilizers, agrichemicals and manures, and so contemporary monitoring of inputs of pollutants is complicated by those extraneous substances which are already present. Given this variability in the background soil contents of a wide range of pollutants, it is essential to sample a sufficiently large number of sites to determine local background concentrations of pollutants for comparison with the more recent or contemporaneously polluted sites. For example, if monitoring is carried out for TCDD in soils near to an incinerator, the soil content of the pollutant may already have been significantly high due to the use of herbicides with traces of the pollutant or the disposal of sewage sludge containing traces of TCDD. This has not always been done because workers have relied on comparisons of the data for suspected polluted sites with national or international average values. If all the soils in an area had previously been polluted or the soil parent material was naturally (geochemically) enriched in some metals (for example Cd, As, Pb, Cu, Zn, Mo and V in marine black shales), then it is not possible to tell whether the suspect soil is polluted or not until the local background levels are known. However, strictly speaking, although it may not be polluted in the sense of Holdgate's definition (chapter 1), the soil still contains anomalously high levels of several heavy metals which may be taken up by food plants and pass along food chains and have similar effects to the same metals arising from pollution rather than geochemical enrichment.

Soil analysis for atmospherically deposited pollutants will normally be based on shallow sampling (0–2.5 or 5 cm) in topsoils. Intentionally placed pollutants which were subsequently ploughed into the soil need to be studied in samples of both the topsoil (normally 0–15 cm, but sometimes 0–10 cm) and samples from deeper in the soil profile (15–30 cm and 30–45 cm). For normal agricultural testing purposes, a composite sample of 25 auger cores of soil in a W-pattern covering about 5 ha are collected and bulked together to give one composite sample. Patches of different soils, such as distinctly wet areas or where the crop shows distinct differences in colour or growth, should be sampled separately. The taking of 25 auger cores helps to overcome the problems of soil heterogeneity. If the bulk samples are too large for logistic reasons, they should be thoroughly mixed and systematically split to obtain smaller representative samples. Soil sample collection, treatment and analysis for heavy metals is comprehensively covered by Ure (1990).

Analytical surveys of contaminated or derelict land differ from many other types of environmental monitoring because they are usually carried out on a one-off basis and are intended to determine whether any pollutants are present in significantly high concentrations which will need special attention in the development of the site. There are many cases in the literature of housing being built on hazardous waste landfill that had not been thoroughly investigated before development (see section 7.6). Site surveys are usually based on a sampling grid going down several metres to the undisturbed subsoil. If the buildings have already been cleared the sample grid should be selected using any site plans or any other information available to ensure that the areas most likely to be severely polluted (such as waste lagoons, chemical stores, etc.) are included in the survey (Finnecy and Pearce, 1986; Alloway, 1992).

Soil samples are usually air-dried, disaggregated and passed through a nylon sieve with 2 mm apertures, and then thoroughly homogenized ready for analysis. All soil analyses require the pollutants to be either extracted with an appropriate solvent in the case of organic compounds, chelating agents, acids or salt solutions for extractable metals, or the soil digested in boiling strong acids for total metal analysis followed by atomic absorption spectrophotometry. Organic pollutants are determined by the chromatographic methods described in sections 4.1–4.4.

Where possible, plant analysis should be based on samples of the same tissue in the same position on the plant (e.g. flag leaf in cereals) at the same height above ground and from plants of the same species at the same stage of growth. In practice this is not always possible but should be strived for wherever it can be carried out. Leaf samples should be washed with deionized distilled water to remove particulates deposited from the atmosphere and soil-splash and then the leaves should be dried in a forced draught oven at 60°C until they reach constant weight when they should be ground to a powder in a mill with no contaminating metal parts in contact with the sample. Analysis of the plant material for metals usually involves either digesting in boiling concentrated HNO_3, or igniting in a furnace, to destroy the organic matter. Some organic pollutant determinations involve leaching macerated tissue with organic solvents and further treatments prior to chromatographic analysis.

Ecosystem target monitoring may require certain animals to be sampled and parts of them analysed, especially in food chain studies. Sometimes this can be done post-mortem on animal corpses found in the study area but trapping, blood sampling and other tests may be required. If soil invertebrates are to be sampled and analysed, care needs to be taken to avoid soil contamination. Where possible zoological assistance should be sought in planning the monitoring and identifying the appropriate species and tissues for sampling and analysis. Monitoring larger animals needs veterinary expertise. Most animal tissue analysis involves destruction of the organic matter usually by digestion in strong acids.

Interpretation of monitoring data often necessitates the plotting of isopleths of pollutants onto maps of the area affected, analysis of the variance of data for control and affected areas and various types of regression analysis to evaluate apparent relationships between parameters.

The reader is referred to Hewitt and Allot (1992) for more detailed coverage of the theory of environmental monitoring and Hewitt (1991) for instrumental methods of analysis.

References   Adams, S. (1992) personal communication.

Ahuja, S. (1992) *Trace and Ultratrace Analysis by HPLC*, Wiley, New York.

Alloway, B. J. (ed) (1990) *Heavy Metals in Soils*, Blackie, Glasgow.

Alloway, B. J. (1992) Chapter 5 in Harrison, R. M. (ed) *Understanding Our Environment* (2nd edition), Royal Society of Chemistry, Cambridge.

Braithwaite, A. and Smith, F. J. (1990) *Chromatographic Methods* (4th edition), Chapman and Hall, London.

Braithwaite, A. and Smith, F. J. (1990) Concentration and determination of trace amounts of chlorinated pesticides in aqueous samples. *Chromatrographia*, **30**, 129–134.

CHROMPAK Ltd. (1992) Hydrocarbon smog, Bilthoven air, May 17, 1990. *J. Chromatog. Sci.*, **30**, 40.

Ebdon, L. (1982) *An Introduction to Atomic Absorption Spectroscopy*, Heyden, London.

Finnecy, E. E. and Pearce, K. W. (1986) Chapter 4 in Harrison, R. M. (ed) *Understanding Our Environment* (1st edition), Royal Society of Chemistry, London.

Gopalan, R. (1977) Ph.D. thesis, University of London.

Harrison, R. M. (1990) Chapter 7 in Harrison, R. M. (ed) *Pollution: causes, effects and control* (2nd edition), Royal Society of Chemistry, Cambridge.

Hewitt, C. N. (ed) (1991) *Instrumental Analysis of Pollutants*, Elsevier, London.

Hewitt, C. N. and Allot, R. (1992) Chapter 7 in Harrison, R. M. (ed) *Understanding Our Environment* (2nd edition), Royal Society of Chemistry, Cambridge.

Holdgate, M. W. (1979) *A Perspective of Environmental Pollution*, Cambridge University Press, Cambridge.

Kuwata, K., Uebori, M. and Yamazaki, Y. (1980) Determination of phenol in polluted air as *p*-nitrobenzenazophenol derivative by reversed phase HPLC. *Anal. Chem.*, **52**, 857–859.

Lajunen, L. H. J. (1992) *Spectochemical Analysis by Atomic Absorption and Emission*, Royal Society of Chemistry, Cambridge.

Lane, D. A. (1982) Mobile mass spectrometry: a new technique for mobile environmental analysis. *Environ. Sci. Technol.*, **16**, 45A–49A.

Possanzini, M. and Di Palo, V. (1990) Improved HPLC determination of aliphatic amines in air by diffusion and derivatisation techniques. *Chromatographia*, **29**, 152–154.

Rastogi, S. C. (1990) A routine GC method for the analysis of CFCs in aerosol cans. *Chromatographia*, **29**, 13–15.

Renberg, L. and Lindstrom, K. (1981) Reversed phase trace enrichment of chlorinated phenols in water. *J. Chromatog.*, **214**, 327–334.

Slavin, W. (1992) A comparison of atomic spectroscopic analytical techniques. *Spec. Intern.*, **4**, 22–27.

Stahl, E. (1969) *Thin Layer Chromatography, a Laboratory Handbook*, Allen & Unwin, London, p. 642.

Suelter, C. H. and Watson, J. T. (eds))1991) Biomedical applications of mass spectrometry, in *Methods of Biochemical Analysis*, Wiley, New York.

SUPELCO Ltd (1990) Temperature programmed separation of mixed hydrocarbons, *J. Chromatog. Sci.*, **28**, GC116.

Takeuchi, T. and Ishii, D. (1981) High performance micro-packed flexible columns in liquid chromatography. *J. Chromatog.*, **213**, 25–32.

Ure, A. M. (1990) Chapter 4 in Alloway, B. J. (ed) *Heavy Metals in Soils*, Blackie, Glasgow.

Further reading

Holland, M. R., Smith, B. L. and Perrett, D. (1987) *A Guide to GLC and HPLC*, Assoc. Clin. Biochem., London.

Karasek, F. W., Hutzinger, O. and Safe, S. (1985) *Mass Spectrometry in Environmental Science*, Plenum, New York.

Sherma, J. and Fried, B. (1990) *Handbook of Thin Layer Chromatography*, Dekker, New York.

Part Two
The Pollutants

Inorganic pollutants 5

5.1.1 Historical

Ozone was first identified in 1840 by Schonbein who described its formation by electrolysis and also when an electric discharge passes through the oxygen of the air. Its peculiar odour is detectable on the London underground and in rooms with photocopiers.

5.1.2 Formation

Electrical discharge Laboratory ozonizers depend on energizing oxygen by passage through a silent discharge of some 10 000 volts, which at 0°C produces a gas stream of about 4% ozone by volume. This low conversion is due to the ready thermal decomposition of the initial product:

$$3\,O_2 \overset{\Delta}{\rightarrow} 2\,O_3 \tag{5.1}$$

If air is passed through an ozonizer then some combination of nitrogen and oxygen occurs to give nitric oxide which is more stable to heat than is ozone. This reaction occurs during thunderstorms and most significantly in the internal combustion engine.

$$N_2 + O_2 \rightarrow 2\,NO \tag{5.2}$$

5.1.3 Physical properties and structure (Advances in Chemistry, 1959)

Ozone has a melting point of −93°C, a boiling point of −112°C and a density of 1.6 relative to air (= 1.0). The bonds linking the oxygen atoms in ozone are 0.128 nm in length and they are disposed at an angle of 116°. The principal structure (**1**) is a charge separated species which arises from the donation of a shared electron pair from one terminal bond to the other.

Absorption of ultraviolet light. Oxygen absorbs visible light and this corresponds to dark Fraunhofer lines in the absorption spectrum of sunlight, notably at 759 nm (extreme red) and 687 nm (red). Activation at

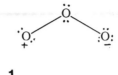

1

these low quantum levels does not lead to chemical change only to dispersal of the energy as heat. However, absorption of higher energy radiation at shorter wavelength, typically at 200 nm, leads to cleavage of the molecule and the formation of very reactive atomic oxygen (equation 5.3). When atomic and molecular oxygen collide ozone is formed and survives the collision provided that a third inert entity (M), normally nitrogen present in excess, is available to take up excess energy.

$$O_2 \xrightarrow{h\nu} [O] + [O] \tag{5.3}$$

$$[O] + O_2 + M \rightarrow O_3 + M \tag{5.4}$$

$$O_3 \xrightarrow{h\nu} O_2 + O \tag{5.5}$$

The activated forms atomic oxygen and ozone are sometimes described as 'odd oxygen'. The principal UV absorption of ozone occurs in the range 200–300 nm and irradiation with light of this wavelength will lead to reversal of ozone formed as in (5.3). Ozone is detectable by absorption spectroscopy at levels down to 0.01 ppm.

5.1.4   The ozone layer

Figure 5.1 shows the distribution of atomic and molecular oxygen and of ozone as a function of distance from the earth's surface. In the distant stratosphere oxygen is at a low concentration and the rate of formation of ozone shown in equation 5.4 is reduced; at the same time the radiation becomes more intense which enhances the decomposition of ozone according to equation 5.5. As a result atomic oxygen is here the dominant form of odd oxygen.

At about 60 km from the earth's surface increasing concentrations of molecular oxygen and nitrogen favour ozone formation (equation 5.4), while a decrease in the incidence of higher energy light disfavours formation of atomic oxygen. As lower levels are reached ozone becomes the only form of odd oxygen and it forms a layer in the region of 35 km where it accumulates at a concentration of about 1% of the atmosphere. The formation of this layer of ozone extends the absorption of UV light to include the range 200–300 nm and these frequencies are filtered from the sun which falls on the earth. Below the ozone layer in the troposphere the undisturbed system is protected from high frequency light and in its absence the formation of atomic oxygen and hence of ozone is minimal.

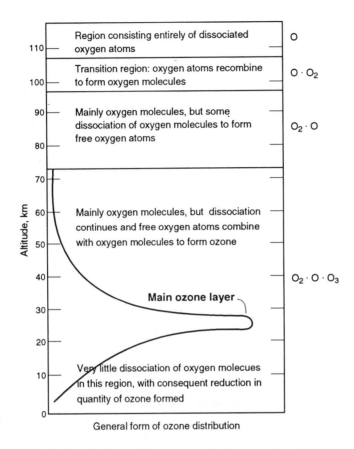

Figure 5.1 Distribution of ozone and its precursors in the stratosphere (Nebel, C., Encyclopedia of Chemical Technology (3rd edn), Copyright © 1981, reprinted by permission of John Wiley & Sons, Inc.).

5.1.5   Factors which disturb the natural environment

Depletion of the ozone layer.   Measurement of ozone concentrations depends on the absorption of UV light already referred to and also to the typical absorption or emission of infrared light. Analysis is complicated by natural seasonal variations and also by changes in the solar emission cycle and by stratospheric winds. Observations have been going on for 25 years, however a combination of ground-based and satellite-based instruments has provided the most telling data of weakening of the layer. A large body of evidence includes a spring depletion up to 30% above Antarctica substantiated by an observed sequential October decline in the seven years to 1987.

Continued depletion of the ozone layer must have serious consequences for the earth's ecology since it would lead directly to an increased exposure to light in the 200–300 nm range. This radiation is toxic to unicellular organisms and to the surface cells of animals and higher plants. An increase in *treatable* human skin cancers has already been confirmed although it is not yet clear whether an increase in the more dangerous melanoma arises in the same way, since there are other environmental factors and they usually occur in parts of the body not directly exposed to sunlight.

Aviation. In addition to contributing to the burden of hydrocarbons, supersonic transports, such as Concorde, operate at a height sufficient to interfere with the protective ozone layer. Nitrogen and oxygen combine to produce nitric oxide at the high temperature of the engines, the ozone level is then reduced by the combination:

$$NO + O_3 \rightarrow NO_2 + O_2 \qquad (5.6)$$

Release of chlorofluorocarbons. These substances were first marketed by General Motors Laboratories in 1930 as they were non-toxic, non-flammable and could replace SO_2, NH_3 and carbon tetrachloride in refrigerators. On the evidence available at the time they appeared ideally friendly to the environment since they were stable to all the reactions which take place in surface air pollution (see p. 31) and are unaffected by normal daylight radiation. They came into increasing use as commercial refrigeration became widespread and also found applications as industrial solvents and blowing agents (see p. 213).

The two substances of significance are trichlorofluoromethane or CFC-11 and dichlorodifluoromethane or CFC-12, where the units refer to the number of fluorine atoms and the ten to the single carbon atom. They tend to persist in the troposphere on account of their chemical stability and because they are volatile by choice, insoluble in water and are not removed by rainout. It is now realized that these properties make them undesirable as 'greenhouse' gases but even more seriously it was shown by Molina and Rowland in 1974 that they diffused upwards into the stratosphere and were there exposed to high frequency radiation which caused them to react and destroy ozone (see equations 5.7–5.9). At that time production of CFC-11 was 0.3 million tons/year and of CFC-12 0.5 million tons/year and it was shown that stratospheric levels then correlated with the total produced up to that time. In subsequent years growth in the USA was almost 9% per year.

5.1.6   Chemistry of stratospheric CFC (Wayne, 1991)

In contrast to their stability in the troposphere the higher energy radiation at 20–40 km causes photolysis:

$$CFCl_3 \xrightarrow{h\nu} CFCl_2 + Cl^{\cdot} \qquad (5.7)$$

$$CF_2Cl_2 \xrightarrow{h\nu} CF_2Cl + Cl^{\cdot} \qquad (5.8)$$

Chlorine atoms produced in this way act to destroy ozone in a chain reaction in an analogous manner to the oxides of nitrogen (see p. 112):

$$Cl^{\cdot} + O_3 \rightarrow ClO^{\cdot} + O_2 \qquad (5.9)$$

As it stands this reaction would have little effect but in the ice clouds which form over Antarctica, chlorine monoxide radicals disproportionate to yield molecular oxygen and chlorine atoms, which re-enter the cycle and can account for very many ozone molecules.

Nomenclature.   For brevity a system is used which is based on the form CFC-*XYZ* where X = number of carbon atoms minus 1 (omitted if $X = 0$); Y = number of hydrogen atoms plus 1; Z = number of fluorine atoms. So that $CFCl_3$ becomes CFC-11 and CF_2Cl_2 is CFC-12.

Ozone depletion potential (ODP).   This is based on model calculations made at the University of Oslo and trichlorofluoromethane (CFC-11), the most damaging chlorofluorocarbon, is assigned an ODP of 1.0. On this scale dichlorodifluoromethane (CFC-12) has an ODP of 0.9. The potential for depletion will be reduced if a substance is less persistent – CFC-11 survives the decade it takes to diffuse into the stratosphere and persists for many years more – however, research is going on to develop alternatives whose concentration is reduced by reaction in air at lower levels. These compounds typically include a reactive C–H bond and are known as HCFCs, they decompose in the troposphere under attack by species such as the $^{\cdot}$OH radical. This initiates change by the abstraction of $H^{\cdot}$ from the C–H bond and the reaction type is similar to that discussed on p. 203.

Examples of suitable compounds are: dichlorotrifluoroethane CF_3CHCl_2 (HCFC-123) for blowing foam (ODP 0.013); tetrafluoroethane CF_3CH_2F (HCFC-134) as a refrigerant (ODP zero); 1,2-dichloro-1,1-difluoroethane $CH_2Cl–CF_2Cl$ (HCFC-132) as solvent in the electronics industry.

Further examples of the environmental exposure of these compounds are given in the section on solvents (pp. 213–4).

5.1.7 Control measures (American Conference of Government Industrial Hygienists, 1990)

Legislation was enacted in the USA to restrict the use of CFCs as early as 1975 and worldwide production declined through the 1980s. In 1987 an International Protocol was agreed at Montreal whereby the signatories agreed to halve their current production by 1999. In recent years the public and governments have become increasingly concerned. An international work group reported 18% depletion of the ozone layer over Europe in the winter of 1991/2 and work in the USA has confirmed this trend. An increase in skin cancer and eye cataracts has been predicted and in the UK the Health Education Authority has issued a warning about over-exposure to the sun.

An international meeting at Copenhagen in November 1992 took account of the growing problem and agreed for the most part to phase out CFCs altogether by 1995. Desirable though this is, it should be noted that considerable usage continues in countries outside the agreement and some manufacturers have been accused of selling stocks to these consumers.

5.1.8   Ozone in the troposphere

The filtering action of the ozone layer removes the shorter wavelength radiation with sufficient energy to produce odd oxygen (p. 110). Under natural conditions small amounts of ozone enter by transfer from the stratosphere. However, tests in pollution zones reveal substantial levels which in sunlit urban areas can rise to 1 μg/g; a typical concentration is about 0.3 μg/g. In these areas nitrogen is fixed as nitric oxide in the internal combustion engine where oxygen is present in limited excess (see equation 5.2). In daylight this initial product reacts with atomic oxygen, or other oxidizing agent present in low concentration in the natural environment, to give the peroxide:

$$NO + O \rightarrow NO_2 \qquad (5.10)$$

This is the primary source of ozone because it can be photolysed by available blue light of wavelength near 400 nm. So through the morning, sunlight initiates decomposition of the peroxide to give the lower oxide and release atomic oxygen:

$$NO_2 \rightarrow NO + [O] \qquad (5.11)$$

This then combines with molecular oxygen to produce ozone.

In the absence of other agents, oxygen and $(NO)_x$ would achieve a steady state including a low level of ozone, but equation 5.11 is driven to the right by the removal of NO through its reaction with other pollutants, notably carbon monoxide and unburnt hydrocarbons (see p. 203). As a result the concentration of atomic oxygen and hence of ozone is further increased.

In this context it must be stressed that the environment is not compartmentalized and although use of HCFCs reduces the damage to the ozone layer they act in the troposphere as greenhouse gases (p. 119).

Methyl bromide is a substance of current concern not so much from its action in the troposphere, but from the realization that its survival in low concentrations in the stratosphere releases very damaging bromine atoms. About 50% of methyl bromide comes from natural marine sources, but the rest is man-made because its low boiling point of 5°C has lead to widespread use as a space and soil fumigant. The quality and variety of foodstuffs available in advanced countries depend on the pesticidal action of this compound. It is used on food crops, coffee, cocoa, forage, timber, tobacco, etc., besides the protection of foods and grain in storage and during transportation in ships, trains and trucks. There is no single substitute although ethyl ether may be used as a fumigant, Malathion (p. 252) for stored products and chloropicrin for soils.

5.1.9   Diurnal variations of ozone levels

In the stratosphere little change occurs during the dark hours since, although production ceases, the concentration of atomic oxygen falls away and ozone is no longer lost by the reaction:

$$O + O_3 \rightarrow 2\,O_2 \qquad\qquad (5.12)$$

In contrast, levels in the troposphere decline after dark because ozone continues to react with other residual pollutants such as hydrocarbons. A typical diurnal cycle is shown in the following chapter (p. 205).

Ozone differs from molecular oxygen in that it is a polarized molecule (see **1**) and hence is twelve times as soluble in water as oxygen under comparable conditions of temperature and pressure: ozone is therefore scrubbed from polluted air by rainfall. Uptake of ozone from ozonizers may produce aqueous solutions with concentrations of the order of 40 mg/l and these are useful in destroying the last traces of pollutants in waste waters (see p. 269).

5.1.10   Toxicity and control

The primary site of injury is the lung which may become congested with swelling of the tissue (oedema) and possible haemorrhage. In man exposure for 2 hours to concentrations of 1.5 ppm is regarded as dangerous. Workers particularly at risk are welders using the shielded arc process. A threshold limit value (TLV) of 0.1 ppm is being considered in Europe and the USA.

5.2
Oxides of carbon, nitrogen and sulphur

The problems of growth associated with demand for energy and supply by combustion of fossil fuels (section 6.2) were stressed in chapter 1. This section covers the chemical and physical properties of the pollutants which are produced. Table 5.1 shows the composition of clean air.

Table 5.1 The composition
of clean air by volume (%)

Nitrogen	78.1
Oxygen	20.9
Argon	0.9
Carbon dioxide	0.034

5.2.1   Carbon dioxide

Variation of yield with source.   It is shown in section 6.2 (p. 200) that most coals contain 85–90% of carbon, almost all of which is released as carbon dioxide on combustion. This must be compared with a typical fuel oil whose composition will be close to that of a saturated aliphatic hydrocarbon, i.e. $C_nH_{(2n+2)}$. Of this group methane (CH_4) has the lowest carbon content of 75%, which rises on ascending the series to 84% for octane (C_8H_{18}). Hence oil, and especially natural gas, contribute less per energy unit to the CO_2 burden than does combustion of a representative coal (Table 5.2).

   While the more volatile petroleum fractions, including gasoline, are to be preferred for the control of CO_2 emissions the release of oxidized minor components, especially nitrogen and sulphur, must also be considered. The consequences of NO_x release are discussed on p. 124 and those of SO_2 on p. 129.

The observed increase in CO_2 levels.   The natural level of 0.03% corresponds to 275 ppm and correlates with that found in ice cores pre-dating the Industrial Revolution and also with a summary of earlier data made by Callender in 1958. The subsequent upward trend (Figure 5.2) has been established remote from local discharges at the Mauna Loa observatory following an initiative by Keeling.

   These observations are supported by others made at the South Pole and in Alaska at Point Barrow, also by samples taken during aircraft flights.

Table 5.2 Relative CO_2 production/
unit of energy* from various fuels

Oil	2.0
Gas	1.45
Coals	2.5

* The quad = 1015 BTU or 1.055 kJ

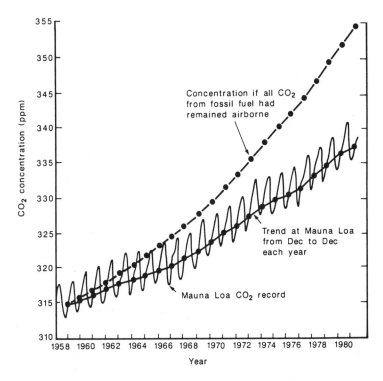

Figure 5.2   Rise in CO_2 concentration in air at Mauna Loa (Keeling, *Geophysics Monograph*, **55**, 165–236, 1989, copyright by the American Geophysical Union).

From Figure 5.2 it can be seen that there is a seasonal downward 'wobble' every year due to the uptake of CO_2 during photosynthesis by deciduous trees and plants. It is also apparent that only part of the released CO_2 remains in the atmosphere. This is to be expected since it is soluble in water to the extent of 3.3 g/l at 0°C and 1.7 g/l at 20°C. Ocean sediments arise from its solution in sea water but the surface contact with natural water is limited and only moderates the rise in concentration.

Ice cores.   The solubility of CO_2 in water ensures that equilibrium is established through time between the levels in air and water. Hence from a knowledge of the level in polar ice of known date it is possible to evaluate the level in air in equilibrium with it at the same period. It has been shown that in the years preceding the Mauna Loa analysis that CO_2 ranged from an unperturbed level of 270 ppm to 320 ppm. Figure 5.3 shows the rise in emission from individual fuels and Figure 5.4 the relative contributions predicted for the year 2025 from major consumer countries.

Wood is an important energy source in Third World communities. For

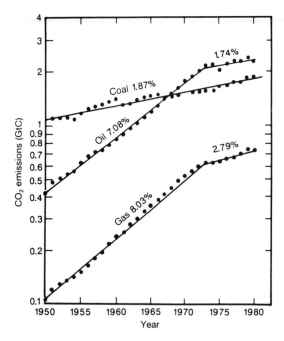

Figure 5.3   Annual emissions of CO_2 by fuel type with indicated growth rates (Rotty, R.M., *J. Geophys. Research*, **88**, 1301, 1983, copyright by the American Geophysical Union).

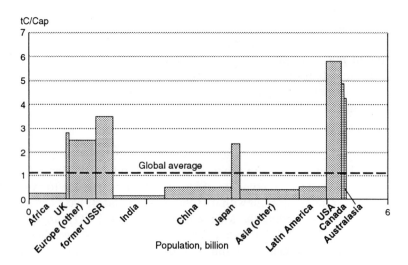

Figure 5.4   Relative carbon emissions per capita and population (1989) (Grubb, 1991).

example in India a village will consume about 3800×10^9 joules annually and of this almost 90% comes from firewood. Taking India as a whole, wood and charcoal account for one-third of its energy demand but this is small on a global scale since the inhabitants consume only one-twentieth of that consumed by inhabitants of developed countries.

By reference to Figure 5.4 and Figure 1.2 it is seen that as heavily populated Third World countries such as India and China strive to raise their living standard, residual CO_2 levels must rise at a greater rate than before. At present they account for 80% of the world's population and consume only 25% of commercial energy. Their demand is expected to double during 1990–2010 and redouble during 2010–2030. After that time if 10 billion people are to meet the desired 60% cut in present levels per capita, emission must fall to one-fifth, i.e. to 0.2 tC/year.

The greenhouse effect (Dickinson, 1986).   Solar energy falling on the earth's surface is absorbed and transferred to the atmosphere by evaporation and as heat flux, including infrared radiation. At present the energy fall at the surface is about 157 W/m^2; rather more than twice this amount enters the stratosphere but is depleted by reflection from clouds and dust and also by absorption by clouds, ozone and water vapour. The infrared energy absorbed by these components is governed by the intensity of infrared emission at the earth's surface and in the lower levels. The emission of energy from the upper levels is reduced as the temperature of the troposphere falls at about 5°C/km, leading to a net energy gain and surface warming.

Man's activities have introduced additional absorptive molecules, so trapping more heat and disturbing the natural equilibrium (Wayne, 1991). The principle components and their efficiencies are shown in Table 5.3.

Figure 5.5 shows the span of the infrared spectrum in the frequency range 500–4000 cm^{-1} which is equivalent to a wavelength range of 20–2.5 μm.

Table 5.3   Trapping of infrared radiation by trace constituents (1985)

	Concentration (ppb)	Trapping (W/m^3)
Carbon dioxide	345×10^3	*c.* 50
Methane	1.7×10^3	1.7
Ozone	10–100	1.3
Nitrous oxide	304	1.3
CFC-11	0.22	0.06
CFC-12	0.38	0.12

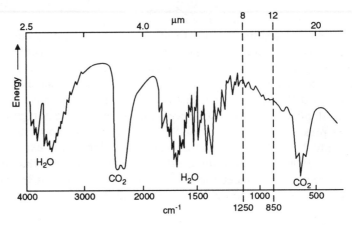

Figure 5.5   Background i.r. absorption by water vapour and CO_2.

The escape of infrared energy is blocked by the absorption of water vapour and carbon dioxide save for a narrow window near 3.5 μm (2800 cm^{-1}) and a broad window in the range 8–12 μm (1250–850 cm^{-1}). It can now be seen why the effect of polluting species is so much greater than is immediately suggested by their relatively low concentration as compared to carbon dioxide (Table 5.3): this is because they have appreciable absorption within the natural escape window. Ozone absorbs strongly at 9.6 μm and at a concentration one thousandth that of carbon dioxide its absorption is as high as one-thirtieth of the major greenhouse gas. The release of ozone and CFCs has already been mentioned (pp. 112, 114), methane is a significant and growing contributor (p. 202) and the contribution of (NO)$_x$ is discussed later in this section.

Effect of greenhouse warming.   A fundamental uncertainty is the past effect of variations in the emissions from the sun whose brightness varies by up to 0.1% in phase with magnetic activity. A change of 0.2–0.5% would account for a fall of 0.4–0.6°C in the Little Ice Age. Future forecasters will benefit from observations by NASA which now records the sun's energy to 0.01%.

It is known from the study of ice cores that CO_2 levels in the past interglacial age were similar to those of today and the conditions on other planets correlate with their surface temperatures. Mars has a thin atmosphere composed mainly of CO_2, a surface pressure less than one-hundredth that of earth and a surface temperature of 223 K. Venus has an atmosphere which is 96% CO_2 at a surface pressure one hundred times that of earth and a surface temperature of 732 K (Wayne, 1991).

The earth's surface temperature has risen by 0.5°C over the last century

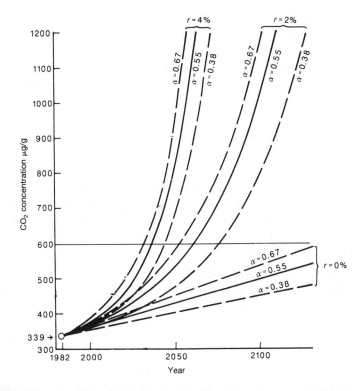

Figure 5.6   Predicted CO_2 levels for zero, 2% and 4% growth (from Roberts *et al.*, 1990).

and predictions depend on forecasting the likely rate of change in the concentration of greenhouse gases. The earth's atmosphere contains 2.6×10^{15} kg of CO_2 to which man added 5×10^{12} kg in 1985 and the current level is 1.9×10^{13} kg/year. Fossil fuel usage over the past century increased at 4.5%/year, a doubling time of 16 years, leading to the present level of 350 µg/g. If the future rate of usage averages 4% then the natural level will be doubled to one of 600 µg/g by the year 2030 (Figure 5.6). With growth nearer to the present rate of 2% this point will not be reached until 2050. These predicted dates will vary somewhat depending on the fraction of emitted CO_2 which persists in the atmosphere: at present this is almost one-half.

Computer models which broadly match conditions in the past indicate a temperature rise of 3°C on doubling present CO_2 levels. This must be qualified as there are both positive feedback mechanisms which enhance the effect and negative feedback which reduces it. One example of negative feedback is the reduction of incident radiation as a result of increased cloud cover as oceans warm up. An example of positive feedback is the reduced reflectivity (albedo) owing to the shrinkage of polar ice. Warming is likely to be higher in the northern hemisphere with its higher level of industrial activity and smaller ocean area. At the second World Climate Conference at Geneva in 1990 a forecast was made of 2–5°C warming by the end of next century.

Aside from CO_2 emissions it must be remembered that increases are to be expected in the concentration of other airborne pollutants especially those listed in Table 5.3.

Consequences of greenhouse warming. Some events which may arise from warming have already been noted:

1. Persistent water shortages in southern England – construction of a desalination plant is being considered by Southern Water.
2. The six warmest years since records were kept all fell in the 1980s.
3. Evidence of warmer water off the Cornish coast comes from the appearance there of the tropical triggerfish and also of red mullet and the marbled electric ray.

Apart from variations in the sun's emissions another phenomenon which may account in part for these observations is 'El Nino', a periodical warming of the Pacific west of Peru which occurred twice in the 1980s. Evidence has recently been obtained of fossilized beech leaves near the South Pole indicating a dramatic warming in the recent geological past.

The effect on sea levels. In 1985 responsible predictions based on computer models were of about 1 metre rise by 2050; these have since been revised downward. In May 1990 the United Nations group on climate change forecast an increase of 6 cm/decade or an average rise of 20 cm by 2030 and 60 cm by 2100.

Levels rise as a result of expansion of the water body as temperatures increase and also of its enhancement by melting South polar ice – melting of North polar ice will not directly change sea level as it is already afloat (a full glass of an iced drink does not spill over if left at room temperature). The North Pole is more vulnerable than the South, which was formed 10 million years earlier, and Arctic ice has been reduced in thickness from about 7 m in 1976 to 4 m in 1987. Hermann Flohn has predicted complete melting of Arctic ice on a temperature rise of 4°C and this is equivalent to a localized increase of 38°C in air temperature, presently −34°C.

The Antarctic ice has contributed to sea levels as average temperatures have increased. There was an abnormally high release of icebergs in 1930 and another similar period is now underway. The British Antarctic Survey assessed the 24th iceberg to be recorded since the mid-1980s; it is moving at 3 miles/day and is of exceptional size at 50 miles in diameter and weighing 10^3 billion tons.

The consequences of melting of ice will not be spread uniformly and the predicted average rise of 20 cm will be seen as one of about 35 cm in Europe. This is due to the more rapid disappearance of Arctic ice and its replacement by a body of darker heat-absorbing water. There exist possibilities of drastic changes as a result of the diversion of ocean currents. Warm Gulf Stream waters could take another course to the detriment of the climate of the British Isles and Europe. The direct threat will be to communities on low-lying islands such as the Maldives, 3.66 m above the level of the Indian ocean. In the South Pacific Tonga, Kiribati and Tuvalu are at risk and observation posts have been established there by the Australian Government.

The backing up of rivers will lead to flooding. The most serious risk is to Bangladesh in the delta of the Brahmaputra. In England the threat is to the fenlands and the estuaries of the Thames, Humber and Severn. In Europe those at risk include the Garonne, Loire, Seine, Rhine and the Elbe.

The effect on agriculture.   Changes in the Arctic referred to above will lead to a progressive shift in the meteorological equator at present at 6°N to within 10–12°N. Although this change may be long delayed the consequences are so serious as to demand our attention.

In the UK the south east will become drier and growth of arable crops would tend towards the west. It is also likely that industry would move to the north-west to secure water supplies. Typical French crops such as maize and sunflowers would grow in the south and the malarial mosquito could breed in these areas of England.

The African desert would move northward and areas of Spain, Italy and Greece could become deserts. Similarly, with reduction of winter rainfall, California would come to resemble Mexican desert and the change would also be seen in the Punjab of India.

Control measures.   In a review by the Brookings Institute (Epstein, 1990) it was proposed that control of emissions could be achieved by allocating levels internationally based on previous consumption. These would be set so that the heavy polluters (e.g. USA, USSR, Japan, UK, Germany) would be restricted and countries in surplus (e.g. India, China, Indonesia, Bangladesh) would be allowed growth. A nation wishing to exceed its quota could only do so by purchasing part of that allocated and unused by another nation with an emission balance in hand. Such a measure would discourage the use of coal which produces more CO_2/energy unit than oil.

The United Nations panel (Houghton, 1990) considers that in order to limit temperature rise to within 0.1–0.2°C/decade the emissions of CO_2, CFCs (p. 112) and NO_x (p. 125) must be reduced by 60% and those of methane by 20%. Set against a history of economic growth and demand for energy by a burgeoning population this is an improbable scenario. However, some desirable measures may be summarized:

1. Energy conservation in building design. In October the CBI stated that 20% of energy was wasted out of an investment of £38 billion.

2. Stricter control of vehicle emissions, more efficient use of fuels through speed limitation, greater use of public transport and introduction of the electric car in cities.

3. Overall increase in forestation – this is likely to make only a limited contribution as an increase of about 10% of all forests over and above felling would be required to take up half of the present CO_2 production of 2.5 Gt carbon/year. The value of this area of suitable land and the demand for its use in food production would compete with planting proposals. Trees planted in areas at the borderline of fertility could become unhealthy if soil warming occurred.

4. Encouraging uptake by algal growth. The National Research Council advised US Congress that dumping iron-rich residues in the Antarctic and off Alaska would make up for an existing growth deficiency in these waters.

5. Taxing inefficient sources such as coal and giving preference to natural gas (p. 116).

6. Removal of CO_2 at power stations by scrubbing emitted gas. To be effective it would be necessary to dispose of the trapped gas out of contact with the atmosphere. Burial in the deep sea after entrapment by and subsequent release from ethanolamine has been proposed. Such processes are themselves energy intensive and could even double the cost of electricity.

7. Wind, wave, hydro-, solar and nuclear power are preferable to option 6. An encouraging report is the construction of a cell which harnesses sunlight with an efficiency of 10%. This is higher than sugar cane which at 7% is the most efficient of the plants. The use of nuclear power (p. 167) is attended by unsatisfactory economics and the risk of exposure to radioactivity.

Monitoring.   The UN panel on climate change has been mentioned; 25 countries are represented on it and it last reported in May 1990. In the UK, contributions are made by the Centre for Agricultural Strategy at Bracknell and by the Institute for Terrestrial Ecology at Abbots Ripton.

The Scripps Institute at La Jolla is attempting to detect warming by placing a source of sound at Heard Island in the southern Indian ocean. Using the boundary between warm and cold water as a wave guide, the time taken to reach remote receiver stations can be measured and will increase perceptibly if ocean warming occurs.

5.2.2   Oxides of nitrogen

Sources.   Nitrous oxide (N_2O) is a stable long-lived gas formed naturally by blue-green algae and by *Rhizobium* bacteria active in the nodules of

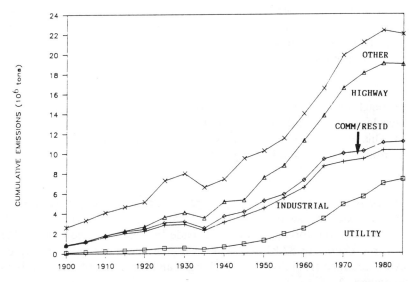

Figure 5.7 NO$_x$ emissions by categories in the USA (COMM/RESID = residential). (From Pahl, D.A., Zimmerman, D., Ryan R., An Overview of Combustion Emissions in the United States, in *Emissions from Combustion Processes*, Clement, R.K., Kagel, R., Lewis Publishers, Boca Raton, Florida, 1990. With permission.)

peas, beans and other legumes. The internal combustion engine, power stations and nitrate fertilizers now produce almost half the amounts arising from the natural sources.

Vehicles are the principal source of nitric oxide (NO) and of its oxidation product, nitrogen peroxide (NO$_2$). The brown peroxide dimerizes to the yellow tetroxide (N$_2$O$_4$) which predominates in highly polluted situations. The analysis of these gases is complicated by their interconversion both in the environment and during chromatography and they are described collectively as NO$_x$.

Levels in air.   In 1987 the concentration of nitrous oxide was 307 ppb and this is increasing by 0.2% annually. Figure 5.7 shows growing contributions to NO$_x$ levels from various anthropogenic sources in the USA, rising in 1985 to 22 $\times$ 10^6 tons.

The levels in other communities can be estimated by scaling these figures in the ratio of their GNP to that of the USA (p. 11).

Effect on the environment.   In addition to the greenhouse effect, NO$_x$ contributes to air pollution in three ways:

1. depletion of the ozone layer;
2. production of acid rain (p. 126);
3. general air pollution.

Depletion of the ozone layer.   Nitrous oxide has a lifetime of about 100 years owing to its low reactivity and once formed naturally or by reduction of surface nitrate it can survive to reach the stratosphere. There it reacts with excited oxygen:

$$N_2O + O \rightarrow 2NO \qquad (5.13)$$

This reaction reduces ozone formation (cf. equation 5.4) and the two molecules of nitric oxide formed lead to ozone decomposition (cf. equation 5.6). This link between excessive use of nitrate fertilizers and depletion of stratospheric ozone could hardly have been predicted.

Production of acid rain (cf. p. 131).   Although oxygen is itself a powerful oxidizing agent it fortunately does not readily combine with nitrogen since the formation of nitric oxide by this route (equation 5.14) has a free energy of 87 kJ/mol. However, the necessary input of energy may be obtained from lightning discharges and from the internal combustion engine.

$$N_2 + O_2 \rightarrow 2NO \qquad (5.14)$$

Further oxidation of nitric oxide by ozone leads to the peroxide, which may then react with hydroxyl or other free radicals to form nitric acid:

$$\cdot OH + NO_2 + M \rightarrow HNO_3 + M \qquad (5.15)$$

This product then falls as rainout contributing significantly to forest decline and with an adverse effect on fisheries, crops and buildings.

Experience in West Germany shows that emissions of NO_x and SO_2 are comparable (Figure 5.8).

The steep rise in NO_x in the years following 1950 are attributable to an increase in vehicle emissions. In the USA over 40% of the NO_x burden arises from vehicles and there it is at a maximum relative to SO_2, which is derived mainly from other sources (cf. Figure 5.8). In 1987 in USA releases of SO_2 and NO_x were comparable at about 20 Tg, six times the West German output. In the UK in 1980 the burden of NO_x from transport was 28% with the major contribution of 46% from power stations. Since the internal combustion engine accounts for only about 3% of released SO_2, a comparison of vehicle usage indicates where measures for control of NO_x are best directed (Table 5.4).

A very high level of private vehicle usage is observed in Los Angeles consistent with the present social threat this represents. The major European cities have a similar pattern of use. Summation of the last two columns gives a transport shortfall for each city which is accounted for by cyclists and pedestrians; this is still significant in Europe and in Amsterdam amounts to 28%. The record for Moscow shows that a major city can function with minimal use of private transport.

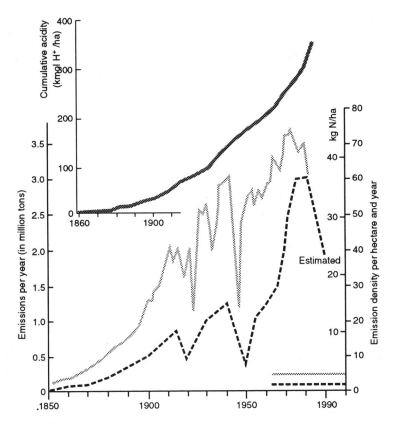

Figure 5.8 Annual emissions of NO$_x$ and SO$_2$ in West Germany (reprinted with permission from Ulrich, B., *Environ. Sci and Technol.*, **24**, p. 439, © 1990 American Chemical Society; Social Trends, 1991).

Table 5.4 Transport patterns in some major cities (1980) (Newman, 1989)

	Petrol use 10³MJ/head	Vehicles/1000 population	Proportion public transport (%)	Proportion private transport (%)
Los Angeles	58	670	7.5	88
New York	44	460	28	64
London	12	355	39	38
Munich	12	400	42	38
Paris	14	385	40	36
Amsterdam	9	340	14	58
Melbourne	29	530	21	74
Moscow	0.4	40	74	2

No_x and air pollution. The critical role of NO_2 as a trigger for tropospheric air pollution has been discussed (p. 114) together with the consequent production of ozone. Nitrogen peroxide also reacts to form peroxyacetylnitrate (PAN) as part of a complex web of reactions that involve methane and other airborne chemicals. A discussion of this chemistry is deferred until the issues arising from the release of organics has been addressed in section 6.2.

Control of NO_x emissions

Emissions from vehicles. It is evident from Table 5.4 that much can be done through economies of consumption of motor fuel and by suitably modifying existing engines. Other possible measures include:

1. Keeping the compression ratio below the usual 10:1. This has the effect of reducing the combustion temperature which is unfavourable to the initial fixation reaction (equation 5.14) and hence reduces the yield of NO.
2. Enriching the mixture of fuel to bring the post-combustion level as low as possible. However, this is energy inefficient and leads to an increase in the release of unburnt fuel.
3. NO and unburnt fuel can be removed from the exhaust using a catalytic converter – the option adopted in North America. The action takes place in two stages. In the first NO_x is reduced to N_2 using unburnt fuel and CO as reducing agents and in the second stage the remaining CO and fuel are oxidized by injection of air at 400°C. The catalysts are platinum or alloys of platinum/rhodium which are poisoned by leaded fuels. SO_2 in the emissions is converted to SO_3 and thence to sulphuric acid (p. 130).
4. Owing to concern at the high levels of emission of NO_x from cars it is likely that all cars sold in the EC after 1992 will have to be fitted with catalytic converters. Table 5.5 shows the relative efficiency of comparable groups of cars, with average engine capacity of 1.7 l, relative to the present EC standard (A) and to that proposed (B).

Table 5.5 Percentage compliance of petrol car emissions with EC standards (Dunne, 1990)

	Present (A)		Post-1992 (B)	
	CO	THC + NO_x	CO	THC + NO_x
EC standard	70%	24%	22%	6%
Uncatalysed	165	81	525	334
Catalysed	10	12	32	47

THC = total hydrocarbons

Table 5.6 Typical sulphur content of petroleum fractions

	% S	Quantity used*
Motor spirit	0.1	24
Diesel fuel	0.3	10.6
Gas oil (domestic)	0.7	8
Fuel oil (power station)	2.0	6.5
Heavy fuel oil (industry)	>3.5**	11

* 1990 in the UK, 10^6 tonnes
** 100 tonnes yield 7 tonnes of SO_2

It is seen that although $THC + NO_x$ is at present within limits, incomplete combustion takes the CO level outside. Furthermore, unless changes are made CO will be five times the future standard and $THC + NO_x$ over three times that permitted.

The future use of electric cars will alleviate the problem by dependence on a power source where NO_x and CO emissions can be brought within legal limits. Citroen have produced a vehicle, the Citela, with a range of 130 miles and a nickel-cadmium battery which can be recharged at home or in parking bays at the rate of 1 mile/minute (*Daily Telegraph*, 1991).

Emissions from power stations.   Nitrogen compounds are present at low concentrations in the fuels and give rise to some NO_x on combustion, but the principal source is fixation of nitrogen by reaction of air in the furnace. Fluidized bed combustion is becoming the standard practice in order to contain emissions of SO_2 (p. 139). These operate at lower temperatures which, as seen above, will also reduce the yield of NO_x.

5.2.3   Oxides of sulphur

Sources

Coal.   The content of sulphur in coal (p. 199) rises to an extreme of 13% in Czechoslovakian supplies but normally varies between 0.5–4.0% with an average amount of about 1.3%; of this about 40% occurs as iron pyrites and the rest is organically combined. There may be significant minor inclusions of chlorine in the range of 0.1–0.7% and traces of fluorine, phosphorus, lead and arsenic.

The biogenetic degradation of natural peptides during the formation of petroleum leads to the inclusion of nitrogen and sulphur compounds. Some of the typical organic components of petroleum, including those containing nitrogen and sulphur, are shown in Figure 6.2; the sulphur content in various fuels is given in Table 5.6. It is also possible that some petroleums are of abiogenic origin and originate in the polymerization of methane which is commonly trapped in sediments and elsewhere in the earth's crust.

Formation and fate.   The worldwide consumption of oil in 1989 was 3.0 billion tonnes with coal consumption at 2.7 billion tonnes oil equivalent (Sprague, 1990). The former was subdivided into low sulphur oil and gasoline (0.5% S), and fuel oil plus heavy oil (2.5% S) yielding some 28 million tonnes of SO_2 in that year. The coal consumed (1.3% S) afforded 39 million tonnes of SO_2. In contrast to CO_2, SO_2 is highly water soluble (23 g/100 g at 0°C; 11 g/100 g at 20°C and NTP) and has a lifetime of only a few hours in the atmosphere before it dissolves in surface water. Its release from vehicles is small compared to that of NO_x but power stations are again implicated being responsible for 60% of the SO_2, twice the quantity produced by industry including the refineries.

Unlike nitrogen, sulphur reacts readily with oxygen during combustion:

$$S + O_2 \rightarrow SO_2 \tag{5.16}$$

$$SO_2 + \tfrac{1}{2}O_2 \rightarrow SO_3 \tag{5.17}$$

The oxidation to sulphur trioxide (equation 5.17) does not occur readily in dry air but proceeds when catalysed by platinum, vanadium or sunlight; ozone is also an effective oxidant (Wayne, 1991). Sulphur dioxide forms sulphuric acid in its reaction with nitric oxide:

$$H_2O + NO + SO_2 \rightarrow H_2SO_4 + N_2O \tag{5.18}$$

but in polluted air the principal route involves the $\cdot$OH radical. In the presence of water and oxygen this can be represented as:

$$\cdot OH + SO_2(+ O_2, H_2O) \rightarrow H_2SO_4 + \cdot OOH \tag{5.19}$$

or more particularly as the initial formation of the bisulphite radical:

$$\cdot OH + SO_2 + M \rightarrow HO.SO_2 + M \tag{5.20}$$

followed by its reaction with water and oxygen:

$$+ \; \cdot OOH \tag{5.21}$$

Effect on the environment.   Sulphur dioxide and the sulphuric acid formed from it have four adverse effects:

1. toxicity to humans;
2. acidification of lakes and surface waters;
3. damage to trees and crops;
4. damage to buildings.

Toxicity to humans (Park, 1987).   The detection limit of SO_2 by humans is about 0.5 μg/g and at levels below 1 μg/g it has no obvious effect but breathing difficulties are experienced above 1.5 μg/g. Exposure at 200 μg/g for 1 minute causes great discomfort, while the limit for prolonged exposure is 5 μg/g. The gas exerts its worst effects on asthmatics and bronchitics who are at risk if exposed for 1 day at the detection limit.

Inhalation of SO_2 was established as the cause of death in a number of incidents:

1930   In the Meuse valley levels over 10 μg/g built up during a thermal inversion with input from power stations, iron and steel works and other industries. Sixty people died on two December days from heart failure linked to respiratory disorder.

1948   Under similar conditions at Donora near Pittsburg over 10 000 people were affected during 5 days in October and 20 deaths were attributed to stress from SO_2.

1952   One of the authors experienced the London 'smog' between 5–9th December which was the worst of a recurrent series of these incidents when to walk outdoors was to travel the unknown. The conditions have been recorded in fiction by Dickens and Conan Doyle. 4000 deaths from lung and heart disease above the normal expectancy were recorded in the months following this worst episode, when SO_2 levels reached 0.7 μg/g. This incident prompted the enactment of the Clean Air Acts of 1956 and 1968 (cf. p. 6).

Acidification of lakes and surface waters.   A measure of acidity is the pH value which is defined as $\log_{10} 1/[H^+]$. The relatively weak carboxylic acids have dissociation constants (K_a) close to 10^{-5}; for acetic acid the figure is $10^{-4.76}$:

$$K_a = [H^+] [AcO^-] / [AcOH] = 10^{-4.76} \qquad (5.22)$$

About 1% of the acetic acid in a 0.1 molar solution is dissociated. From

Table 5.7   Some typical pH values

Strong alkali	14.0	Pure rain	5.7
Sea water	8.5	Acid rain	4.0–4.4
Blood	8.0	Wine	4.0
Conductivity water	7.0	Lemon juice	2.3

equation 5.22 $c = [AcOH]$ and $[H^+] = [AcO^-]$, hence $[H^+]^2 = K_a \times c = 10^{-4.76} \times 10^{-1} = 10^{-5.76}$. Therefore $[H^+] = 10^{-2.88}$ and the pH of a 0.1 molar solution of acetic acid is 2.88.

Decrease of pH can be prevented by buffer action which is best illustrated by a solution containing a salt of a weak base, for example sodium acetate which captures added protons to release the largely undissociated weak acetic acid:

$$Na^+ \, AcO^- + H^+ \rightarrow Na^+ + AcOH \qquad (5.23)$$

For strong acids the dissociation is essentially complete so that $[H^+] = 1.0$ and the pH = log 1.0 = 0. Some typical pH values are shown in Table 5.7.

Many scientists would accept a value less than 5.7 for unpolluted rain, but few would accept the figure of 5.0 proposed by the former CEGB which reveals a vested interest. Because the pH scale is logarithmic it tends to conceal the fact that rain of pH 5.0 has a hydrogen ion concentration 5 times (log 0.7) that of rain of pH 5.7.

Low risk areas are those remote from industrial sources or those which have alkaline soils which 'buffer' and sustain a higher soil pH. Acid lakes can be improved by seeding them with lime but this requires renewal and is expensive.

High risk areas are those which are close to industrial sources or which are exposed to transported acids and the worst consequences appear when the fallout occurs onto thin soils lying on granitic bedrock. Surface waters on granites are not buffered and reveal the trend before soils and plants. A notable consequence is the decline in fish populations where the critical pH is 6.0; Figure 5.9 shows changes in the acidity of lakes in Sweden and New York State during 1930–1975.

Natural buffers in rivers include phosphate, amino acids, bicarbonate and organic matter and their pH depends on an input from soil and rocks in the gathering ground, rather than from direct rainfall on the relatively small area of the river itself. The pH can be affected by changes in land use. A sequence of changes is identified in Table 5.8.

It is now established beyond doubt that atmospheric pollution is linked to the acidification of surface waters. A significant observation is that of Battarbee *et al.* (1985) who showed that trace metals unique to man-made emissions correlated with surface acidity.

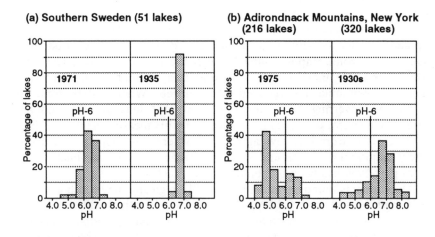

Figure 5.9   Acidification of lakes, 1930–1975 (Wright, 1976).

Table 5.8   Changes consequent on a fall in the pH of surface waters

pH 6.0	Number and variety of species begin to decline
pH <5.8	Green algae and diatoms disappear; more light penetrates, the water is clearer
pH <5.5	Growth of sphagnum moss and filamentous algae is enhanced; matting on lake floor; oxygen access reduced with accumulation of organic matter; bacteria no longer survive

Effect on fish stocks.   Roach, trout and perch cannot tolerate acid waters and many upland lakes such as those monitored for Figure 5.9 have no fish left in them. Depletion of fish populations has been recorded in Scotland, notably in Galloway, and the headwaters of many rivers elsewhere show signs of a decline.

In acid waters the solubilization of heavy metals (p. 140) such as manganese, cadmium, mercury and especially aluminium is the principal cause of fish death. The latter is liberated from clays at pH 4.0 and is toxic to fish even when the pH of the water is not at a harmful level.

Aluminium interferes with reproduction and also damages the gills, while precipitation of aluminium phosphate reduces the primary production of plants and phytoplankton so limiting the food supply.

Effect on soils and vegetation.   Soils usually contain buffering components and hence are much more acid tolerant than surface waters and the practice of liming may give additional protection. A better guide than pH is the *buffer capacity* which is defined as: the number of mols of H^+ or OH^- ions required to lower or raise the pH of 1 kg of soil by 1 pH unit.

Tree growth leads to soil acidification as Ca^{2+} and Mg^{2+} ions are taken up and in order to maintain a charge balance H^+ ions are released. Acid soils may become impoverished through the leaching of K^+, Ca^{2+} and Mg^{2+} ions and by inhibition of the natural decay of litter, leading in turn to increased ammonia and reduced nitrate levels. When soil bacteria are active in converting NH_4^+ ions to nitrate H^+ ions are released and represent a source of acidity not indicated by pH measurements; this is an adverse effect from intensive agriculture, where it is common practice to apply ammonium nitrate at 150 kg of nitrogen/ha. The die-back of ash trees post-1950 has been attributed to this practice.

Aluminium is found in rain at about 0.012 mg/l and may reach a level of 0.45 mg/l in soil and 0.14 mg/l in streams. Relatively little sulphate deposited on soil reaches streams but at pH < 4.0 Al^{3+} is mobilized and its ability to absorb SO_4^{2-} then decreases. The role of Al^{3+} has been disputed but Ulrich (1980) has established that it has a direct effect in damaging fine root hairs and so reducing the uptake of nutrients.

Vegetation damage was first observed in 'desert' areas close to point sources such as smelters, where pine trees are often absent within a radius of 20 km (see p. 9). Lichen deserts also point to acid pollution; these flat plants are especially susceptible as they have no roots and depend on nutrients in rain. Trees and higher plants are affected (a) through denial of nutrients to the roots and (b) damage to exposed foliage.

The 1970s saw evidence of inexplicable damage to coniferous and deciduous trees in the Black Forest of West Germany, which was exposed early on owing to its location close to major industrial output, and in 1980 Ulrich attributed the wasting of its trees to acid rain and defined an unnatural sequence:

1. deposition of nitrate and NO_x in soil accelerates growth;
2. the pH falls and aluminium is mobilized;
3. at pH 4.2 destruction of the roots starts the final deterioration.

Regeneration is difficult on inhospitable acid soils and by 1984 extensive damage to spruce and silver fir to a value of 10 billion DM was reported. Wasting of trees was subsequently identified in the rest of Europe, North America and Canada, which has an extensive forest industry.

Table 5.9 (Innes, 1987) shows the heavy damage to both coniferous and deciduous trees in West Germany, with almost 50% of silver firs in the 26–60% needle/leaf loss group. In 1986 up to 25% loss was occurring right across Europe.

In the UK the Forestry Commission was initially dismissive, attributing the damage to successive drought years. It is now accepted that the observed damage is the result of several factors but that acid deposition contributes significantly to forest decline. The Forestry Commission is

Table 5.9 Forest damage assessment in five European countries (Innes, 1987)

	Needle/leaf loss															
	0–10%				11–25%				26–60%				61–100&			
	83	84	85	86	83	84	85	86	83	84	85	86	83	84	85	86
United Kingdom																
Sitka spruce		65	83	45		28	12	39		6	5	15		1	0	1
Norway spruce		71	84	32		26	15	36		3	1	31		1	0	1
Scots pine		49	74	25		29	18	41		16	7	32		5	1	3
West Germany																
Norway spruce	59	49	48	46	30	31	28	32	10	19	21	20	1	2	3	2
Pine	56	41	43	46	32	38	41	40	10	20	15	13	1	1	2	1
Silver fir	25	13	13	18	27	29	21	22	41	45	50	49	8	13	16	11
Beech	74	50	46	40	22	39	40	41	4	11	13	18	0	1	1	1
Oak	85	57	45	39	13	35	39	41	2	9	16	19	0	0	1	1
Other trees	83	69	69	65	9	24	23	25	8	7	7	9	0	1	1	1
Switzerland																
Norway spruce		65	63	50		28	29	36		6	6	12		1	2	2
Pine		50	35	34		31	47	43		16	13	19		1	5	4
Silver fir		62	60	47		27	28	36		9	8	13		2	4	4
Larch		64	66	39		28	23	44		7	7	12		1	4	5
Beech		74	69	52		23	27	40		3	3	7		0	1	1
Oak		71	60	37		28	33	50		1	6	11		0	1	2
Maple		86	86	73		11	11	25		2	1	1		1	2	1
Ash		84	77	57		13	20	36		3	2	7		0	1	7
Netherlands																
Norway spruce		62	48	49		28	41	34		7	9	12		3	2	4
Scots pine		34	48	50		51	36	33		12	14	13		2	2	3
Corsican pine		57	40	19		34	42	29		8	15	40		1	3	12
Douglas fir		50	33	17		39	43	27		9	22	45		2	2	11
Beech		71	72	68		24	21	26		4	6	5		1	1	2
Oak		57	40	30		38	39	42		5	19	20		1	2	9
Luxembourg																
Norway spruce		79	84	87		17	12	10		3	3	2		2	1	1
Oak		59	77	81		34	20	16		6	3	2		2	1	0
Beech		66	70	67		28	28	27		5	5	6		1	1	1

Table 5.10 Classification of damage to trees (Innes, 1990)

Class	% Needle loss
0 healthy	0
1 slight damage	11–25
2 medium–serious	26–60
3 dying	61–99
4 dead	100

monitoring 141 plots including sitka spruce, Norway spruce and Scots pine and classifies damage in five groups (Table 5.10).

A French study (Leclercq, 1988) based on boring tree trunks and evaluating annual growth cast doubt on the theory that only natural conditions were responsible for decline. In particular, the annual growth of spruce in the Vosges and of silver fir in Luchon did not correspond with the stress events of drought in 1921, 1948 and 1976, nor with the severe frost of 1956.

There is no short term link with SO_2 levels. In Hesse, damage to trees occurred at daily average levels of 800–1600 µg/m, whereas in Freiburg damage was observed at an annual mean level of only 12 µg/m. This must be significant despite the different methods of averaging. Direct damage to trees may occur in fog and cloud water which can have twenty times the concentration of rain water and a pH in the range of 2.8–3.1.

It is difficult to define the effect of individual factors because a considerable time usually elapses between exposure and the emergence of visible signs such as discoloration, shortening of needles and shoots, dry buds and crown die-back. Current thinking is that ozone (p. 109) is implicated but also that NO_x and/or SO_2 may act synergistically and that long term fumigation tests of these agents are needed. A recent study (Spence, 1990) which employed ^{11}C labelling in a test cabinet promises a way to test the toxicity of individual pollutants.

A note on units. It is important when levels of air pollution are measured that the prevailing temperature and pressure of the air body are recorded. This is very necessary for SO_2 levels in mountain forests because a given mass of a gas will occupy a greater volume under the lower air pressure. It is better to quote levels in µg/g (vol./million vols) rather than in mass (mg)/volume (m^3). A table for conversion of these units is given in the appendix; it is based on the following calculation: The ideal gas law states: $pv = nRT$ where p is gas pressure, v the volume, and T the absolute temperature (K). The gas constant $R = 0.082$ l-atm/mol and n = no. of moles of gas. Then $n = W$ (weight)/M (mol.wt) and the gas law can be written as:

$$W/v = M.p/RT \qquad (5.24)$$

To compare µg/g in volume (v)/10^6 volumes of air (V), make the substitution for $V = v.10^6$ in equation 5.24:

$$W/V = M \times p \times 10^6 / R \times T \text{ µg/g} \qquad (5.25)$$

Weight in g must be expressed in mg(10^3)
Volume in l must be expressed in cm^3(10^{-3})

When $W.10^3/V.10^{-3} = M \times 1(\text{atm}) \times 10^6 / 8.2 \times 10^{-2} \times T$ µg/g

So for SO_2 at 20°C = 293 K:

$$W/V = 64 \times 10^6 \times 10^{-3} / 8.2 \times 10^{-2} \times 293 \times 10^3 \text{ µg/g}$$

$$W/V(\text{mg/m}^3) = 2.66 \text{ µg/g} \qquad (5.26)$$

Damage to buildings. In the UK this has been the subject of detailed analysis by the Building Effects Review Group. The buildings worst affected are those finished with limestone which is freely soluble (equation 5.29). St Paul's cathedral is exposed to acid rain at pH 4.0 (Table 5.7) and this pH has been shown to rise due to solution of the building stone, which suffers surface erosion at the rate of 200 μm/year. Damage to the limestone of the colleges of Oxford University cost £7 million to repair in 1991.

Rusting of iron and steel is primarily caused by oxygen and water but it is accelerated by SO_2, nitric acid and chlorides.

Studies of these effects are ongoing through the National Materials Exposure Programme. At the time of the London smog the Beaver Committee estimated losses at £250m. A review of stone weathering in south-east England is available (Janes, 1987).

Monitoring and control. In the UK the Warren Spring Laboratory records acid deposition at 1160 sites and annual records are available (3rd report, 1990). Control will be the responsibility of HM Inspectorate of Pollution, recently established to incorporate the existing control bodies – The Radiochemical, Hazardous Wastes and Industrial Air Pollution Inspectorates with the National Rivers Authority.

Legislative requirements. Control of SO_2 and NO_x are best considered together as they depend on common procedures and set limits for emissions. In Europe for the future these must conform to the EC Framework Directive (84/360/EEC, Speakman, 1990) which defines emission limits to be achieved in 3 phases by the year 2003 (Table 5.11).

The above limits apply only to power stations and therefore exclude about one-third of the emitted SO_2 (p. 130) and half of the NO_x for which motor vehicles are largely responsible (see p. 126).

Table 5.11 EC Targets for major producers of SO_2 and NO_x (Speakman, 1990)

	1980 SO_2 K tonnes	Ceiling K tonnes	Reduction %	1980 NO_x K tonnes	Ceiling K tonnes	Reduction %
Germany	2225	668	−70	870	522	−40
Spain	2290	1440	−50	366	277	−40
France	1910	573	−70	400	240	−40
Italy	2450	900	−70	580	428	−40
UK	3883	1553	−60	1016	711	−30

Control measures.   Precombustion removal of sulphur is chemically feasible for oil through the pressurized hydrogenation of the fuel and alkaline trapping of liberated hydrogen sulphide:

$$R - S - R + 2H_2 \rightarrow 2R - H + H_2S \qquad (5.27)$$

However, the costs are high and are estimated to add 20% to the price of electricity (Tolba, 1983).

There has been considerable research into the cleaning of coal and water washing dissolves part of the iron pyrites (p. 129) and can reduce SO_2 emissions by 10%. The chemical methods largely depend on oxidation and subsequent neutralization of sulphuric acid with lime:

$$4FeS_2 + 15O_2 + 8H_2O \rightarrow 2Fe_2O_3 + 8H_2SO_4 \qquad (5.28)$$

The aqueous slurry is pressurized with oxygen in an autoclave and heated to 150–200°C when most of the inorganic sulphur is removed (equation 5.28).

Organic sulphur compounds (Figure 6.2 p. 201) include thiols (RSH) which react completely under these conditions and some alkyl thioethers (RSR) which react in part, but of the remaining known compounds, aryl ethers (PhSPh) and sulphur heterocycles are unaffected. An added disadvantage is the loss of heating value of the order of 10–20% for only 10–20% sulphur removal. It is therefore policy to control emissions by flue gas desulphurization.

Flue gas desulphurization (FGD).   This was pioneered in the Battersea power station which achieved 80% reduction by scrubbing with water from the river Thames. Figure 5.10 outlines the material flows for the 4000 MW

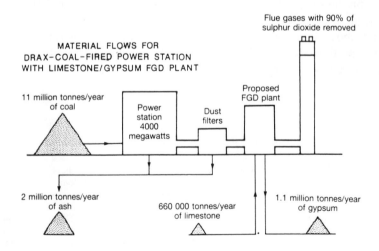

Figure 5.10 Material flows for Drax power station (from Roberts *et al.*, 1990)

Drax power stations. The scrubbing plant injects limestone slurry which converts SO_2 into calcium sulphite:

$$CaCO_3 + SO_2 \rightarrow CaSO_3 + CO_2 \qquad (5.29)$$

and in the same process air is also injected to produce the more tractable calcium sulphate (gypsum):

$$CaSO_3 + \tfrac{1}{2}O_2 \rightarrow CaSO_4 \qquad (5.30)$$

Major new environmental problems arise since extraction of limestone on this scale risks despoilation of the nearby National Park and movement of limestone, gypsum and ash entails some 200 thousand lorry movements a year.

Fluidized bed combustion (FBC).   This technique was researched by British Coal and again depends upon the capture of SO_2 by lime or limestone to form calcium sulphate. Coal in the furnace bed is mixed with limestone and fluidized by passage of air (cf. equation 5.30) at 800–1000°C. A higher temperature must be avoided for two reasons:

1. Calcium sulphate begins to decompose and release SO_2:

$$CaSO_4 \xrightarrow{c.\ 1200°C} CaO + SO_2 + \tfrac{1}{2}O_2 \qquad (5.31)$$

2. In the region of 1400°C thermal fixation of nitrogen occurs to produce NO_x; this may be 40% of the total.

Control of NO_x.   For fuel oils, which have little combined nitrogen, the emission is governed by the furnace temperature and can be kept as low as 200 mg/m^3 by operating in the lower region.

As seen above, FBC can prevent the fixation of nitrogen during coal burning, but combined nitrogen is in the range of 1–2% and will decompose to release ammonia and other nitrogen compounds which are subsequently oxidized to NO_x. If fuel-rich conditions are established in the early stages these substances are reduced to release nitrogen which escapes at the lower operating temperature, reducing the NO_x emission by up to 30%.

Coal volatiles will reduce NO:

$$2NO + 2C \rightarrow N_2 + 2CO \qquad (5.32)$$

and if further control is needed other reducing agents such as ammonia can be injected:

$$6NO + NH_3 \xrightarrow{c.\ 900°C} 5N_2 + 6H_2O \qquad (5.33)$$

An FBC plant can process coal and keep NO_x emissions below the EC requirement of 650 mg/m^3.

5.3.1   General properties

'Heavy metals' is a general collective term applying to the group of metals and metalloids with an atomic density greater than 6 g/cm. Although it is only a loosely defined term it is widely recognized and usually applied to the elements such as Cd, Cr, Cu, Hg, Ni, Pb and Zn which are commonly associated with pollution and toxicity problems. An alternative (and theoretically more acceptable) name for this group of elements is 'trace metals' but it is not as widely used. Unlike most organic pollutants, such as organohalides, heavy metals occur naturally in rock-forming and ore minerals and so there is a range of normal background concentrations of these elements in soils, sediments, waters and living organisms. Pollution gives rise to anomalously high concentrations of the metals relative to the normal background levels; therefore, presence of the metal is insufficient evidence of pollution, the relative concentration is all important.

Apart from aerosols in the atmosphere and direct effluent discharges into waters, the concentrations of heavy metals available to terrestrial, aquatic and marine organisms (i.e. their bioavailability) is determined by the solubilization and release of metals from rock-forming minerals and the adsorption and precipitation reactions which occur in soils and sediments. The extent to which metals are adsorbed depends on the properties of the metal concerned (valency, radius, degree of hydration and coordination with oxygen), the physico-chemical environment (pH and redox status), the nature of the adsorbent (permanent and pH-dependent charge, complex-forming ligands), other metals present and their concentrations, and the presence of soluble ligands in the surrounding fluids.

Although heavy metals differ widely in their chemical properties they are used widely in electronics, machines and the artefacts of everyday life as well as 'high-tech' applications. Consequently they tend to reach the environment from a vast array of anthropogenic sources as well as natural geochemical processes. Some of the oldest cases of environmental pollution in the world are due to heavy metal use such as Cu, Hg and Pb mining, smelting and utilization by ancient civilizations, such as the Romans and the Phoenicians.

The primary production of heavy metals in 1930 and 1985 and the global emissions of metals to soils in the 1980s are shown in Table 5.12.

The data in Table 5.12 show that production of all the metals has increased over the 55 year period. Nickel showed the greatest increase (×35), followed by Cr (×17) and Cd (×14), whereas that for Hg production had not even doubled. The greater emission to soil than annual production for Cd and Hg is probably explained by the occurrence of Cd as

Table 5.12 Primary production of metals and
global emissions to soil (10^3t/yr) (from Nriagu, 1988)

Metal	Production in 1930	Production in 1985	Global emissions to soil in 1980s
Cd	1.3	19	22
Cr	560	9940	896
Cu	1611	8114	954
Hg	3.8	6.8	8.3
Ni	22	778	325
Pb	1696	3077	796
Zn	1394	6024	1372

a contaminant of fertilizers and other metal ores and Hg being both a
constituent of other metal ores and also being emitted naturally from
volcanoes.

5.3.2   Biochemical properties of heavy metals

Some of the elements in this group are required by most living organisms in
small but critical concentrations for normally healthy growth (referred to
as 'micronutrients' or 'essential trace elements') but excess concentrations
cause toxicity. Those metals which are unequivocally essential, whose
deficiency causes disease under normal living conditions include Cu, Mn,
Fe, and Zn for both plants and animals, Co, Cr, Se and I for animals and
B, Mo for plants. Most of the micronutrients owe their essentiality to being
constituents of enzymes and other important proteins involved in key
metabolic pathways. Hence, a deficient supply of the micronutrient will
result in a shortage of the enzyme which leads to metabolic dysfunction
causing disease.

   Some other elements have been shown to have some beneficial effect
under rigorous experimental conditions but are not likely to be responsible
for deficiency disorder under normal conditions. Elements with no known
essential biochemical function are called 'non-essential elements' but are
sometimes also referred to (incorrectly) as 'toxic' elements. These elements, which include As, Cd, Hg, Pb, Pu, Sb, Tl and U, cause toxicity at
concentrations which exceed the tolerance of the organism but do not
cause deficiency disorders at low concentrations like micronutrients. These
are clearly shown by the typical dose response curves in Figure 5.11.

   At the biochemical level, the toxic effects caused by excess concentrations of these metals include competition for sites with essential
metabolites, replacement of essential ions, reactions with $-SH$ groups,
damage to cell membranes, and reactions with the phosphate groups of
ADP and ATP. Organisms have homeostatic mechanisms which enable

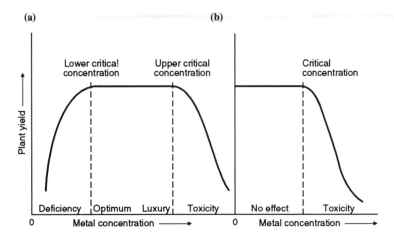

Figure 5.11 Typical dose-response curve for (a) micronutrients and (b) non-essential trace elements (adapted from Alloway (1990)).

them to tolerate small fluctuations in the supply of most elements but prolonged excesses eventually exceed the capacity of the homeostatic system to cope and toxicity occurs, which if severe can cause the death of organisms. An example of homeostasis in animals and the control of excess metals is the formation of metallothionein proteins containing $-SH$ groups which bind certain metals, such as Cd and Zn and enable them to be excreted without causing biochemical dysfunction. In plants, similar compounds called phytochelatins carry out the same function binding divalent metals, such as Cd, in physiologically inactive forms.

5.3.3   Sources of heavy metals

Geochemical sources.   In geological terms heavy metals are included in the group of elements referred to as 'trace elements' which together constitute less than 1% of the rocks in the earth's crust; the macroelements (O, Si, Al, Fe, Ca, Na, K, Mg, Ti, H, P and S) comprise 99% of the earth's crust. These trace elements occur as 'impurities' isomorphously substituted for various macroelement constituents of the crystal lattice of many primary minerals. Primary minerals are those found in igneous rocks which originally crystallized from molten magma. In sedimentary rocks trace elements occur sorbed to the secondary minerals which are the products of the weathering (physical disintegration and chemical decomposition) of primary minerals.

Primary and secondary minerals differ widely in their trace element content and hence the igneous and sedimentary rocks which they respect-

Table 5.13 Typical concentrations of heavy metals in the earth's crust and major types of igneous and sedimentary rocks (from Alloway (1990) based mainly on Krauskopf (1967) and Rose *et al.* (1979))

	Earth's crust	Igneous rocks			Sedimentary rocks		
		Ultramafic	Mafic	Granitic	Limestone	Sandstone	Shales/clays
Ag	0.07	0.06	0.01	0.04	0.12	0.25	0.07
As	1.5	1	1.5	1.5	1	1	13 (<900)
Au	0.004	0.003	0.003	0.002	0.002	0.003	0.0025
Cd	0.1	0.12	0.13	0.09	0.028	0.05	0.22 (<240)
Co	20	110	35	1	0.1	0.3	19
Cr	100	2980	200	4	11	35	39
Cu	50	42	90	13	5.5	30	39
Hg	0.005	0.004	0.01	0.08	0.16	0.29	0.18
Mn	950	1040	1500	400	620	460	850
Mo	1.5	0.3	1	2	0.16	0.2	2.6
Ni	80	2000	150	0.5	7	9	68
Pb	14	14	3	24	5.7	10	23
Sb	0.2	0.1	0.2	0.2	0.3	0.05	1.5
Se	0.05	0.13	0.05	0.05	0.03	0.01	0.5
Sn	2.2	0.5	1.5	3.5	0.5	0.5	6
Tl	0.6	0.0005	0.08	1.1	0.14	0.36	1.2
U	2.4	0.03	0.43	4.4	2.2	0.45	3.7
V	160	40	250	72	45	20	130
W	1	0.1	0.36	1.5	0.56	1.6	1.9
Zn	75	58	100	52	20	30	120

ively form also show a wide variation in heavy metal content. Typical ranges of values for heavy metal concentrations in the earth's crust and various major rock types are given in Table 5.13.

In view of the high degree of variation in the metal contents of rocks shown in Table 5.13, there is a possibility that the soils and stream sediments in a locality suspected of being polluted may have developed from rocks with anomalously high concentrations of certain heavy metals and that pollution, in the strict sense of the definition, has not occurred. Nevertheless, the natural enrichment of metals in the soils may still give rise to harmful effects in living organisms. The natural members of an ecosystem in an area of geochemical enrichment will have evolved tolerance to the elevated concentrations of metals, but newly introduced plant and animal species may be adversely affected. It is therefore important to determine the local background concentrations of heavy metals in order to determine whether the concentrations in the soils and sediments under investigation are significantly higher than those of the area. The normal procedure is to conduct a survey of at least 15 to 20 samples from over the area (the number depending on the size of the area and the degree of heterogeneity in soil types) and to determine the arithmetic mean concentration and its standard deviation (SD) for the range of elements being investigated. Anomalously high (or low) concentrations will lie outside the value for the mean $\pm$ 3 SDs, but values of 2 SDs above the mean are regarded as 'threshold' values.

Heavy metals present in the atmosphere and the hydrosphere.   Table 5.14 gives the concentrations of metals in the atmosphere in three types of location, and the ranges reported in fresh water and sea water. The concentration of heavy metals in the atmosphere in an area remote from anthropogenic effects (South Pole) are markedly lower than those from various locations in Europe, with its range of industrial and urban pollution, as well as relatively remote rural areas. The maximum concentrations found in the US are quoted for some elements to indicate the highest concentrations monitored for technologically advanced countries. In the case of the chalcophyllic elements (those normally found as sulphides in ore bodies, such as Cd, Cu, Pb and Zn) the concentrations were particularly high near volcanoes (Etna and Hawaii) as a result of the lavas containing high concentrations of these elements. The freshwater

Table 5.14   Metals in the atmosphere and hydrosphere (from Bowen, 1979)

Metal	Atmosphere (μg/l)			Hydrosphere (μg/l)*	
	South pole	Europe	Near volcano	Fresh water	Sea water
Ag	<0.0004	0.2–7	3	0.001–3.5	0.03–2.7
As	0.007	1.5–53	<850	0.2–230	0.5–3.7
B	—	3.5	680	7–500	av 4.44k
Be	—	0.9–4	—	0.01–1	2–63
Cd	<0.015	0.5–620	8–92	0.01–3	<0.01–4
Cr	0.005	1–140	45–67	0.1–6	0.2–50
Cu	0.036	8–4.9k	200–3k	2–30	0.05–12
F	—	1.5 (US<400)	high	50–2.7k	av 1.3k
Fe	0.84	130–5.9k	1k–10k	10–1.4k	0.03–70
Hg	—	<0.009–2.8	18–250	0.0001–2.8	0.01–0.22
Mn	0.01	9–210	55–1.3k	0.02–130	0.03–21
Mo	—	<0.2–3.2	—	0.03–10	4–10
Ni	—	4–120	330	0.02–27	0.13–43
Pb	0.63	55–340 (US<13k)	28–1200	0.06–120	0.03–13
Sb	0.0008	0.6–3.2	45	0.01–5	0.18–5.6
Se	0.0056	0.15–11	<2100	0.02–1	0.052–0.2
Sn	—	1.5–800	—	0.0004–0.09	0.002–0.81
U	—	0.2 (US<0.5)	—	0.02–5	0.04–6
V	0.0015	5–92 (US<2k)	79	0.01–20	0.09–2.5
W	0.0015	0.35–1.5	—	<0.02–0.1	0.9–2.5
Zn	0.03	13–16k	10k	0.2–100	0.2–48

* μg/l is equivalent to ppb

samples generally demonstrated a wider range than sea water owing to the effect of salinity in controlling solubility products.

Anthropogenic sources.   Although heavy metals are ubiquitous in most natural materials, the following are significant sources of metals to the environment.

Metalliferous mining. The metals utilized in manufacturing are obtained from either the mining of ore bodies in the rocks of the earth's crust, or the recycling of scrap metal originally derived from geological sources. Ore bodies are naturally occurring concentrations of minerals with a sufficiently high concentration of metals to render them economically worthwhile exploiting. With increasing demand, generally rising prices and improvements in the technology of mineral extraction there is a trend for orebodies with progressively lower metal contents to be used. Generally, these lower grade orebodies are larger in extent and require a higher proportion of rock to be mined per tonne of metal extracted. This inevitably implies that the environmental impact of the mining operations is greater than for mines with smaller areas of much higher grade ore. There will be a need to dispose of greater amounts of tailings, which are the finely milled fragments of rock and some ore particles left behind after the extraction of the metal ore concentrates by various means based on density, magnetism and surface tension (froth flotation). Tailings disposal from current mining operations and continued weathering (chemical alteration) of ore minerals in historical and abandoned mining sites is an important source of heavy metals into the environment (see Figure 7.1). Some of the most common ore minerals of non-ferrous metals are shown in Table 5.15.

From this table it can be seen that many of the major metalliferous ores are sulphide minerals and this has several environmental implications. Fragments of these minerals in tailings deposits oxidize on weathering to create acidic solutions which tend to decrease adsorption and hence increase the mobility of the metals in soils, sediments and waters. The oxidation of pyrite (FeS_2) is shown in detail below:

$$FeS_2 + \tfrac{7}{2}O_2 + H_2O \rightarrow Fe^{2+} + 2SO_4^{2-} + 2H^+ \qquad (5.34)$$

$$Fe^{2+} + \tfrac{5}{2}H_2O + \tfrac{1}{4}O_2 \rightarrow Fe(OH)_3 \text{ (solid)} + 2H^+ \qquad (5.35)$$

$$Fe^{2+} + \tfrac{1}{2}O_2 + H^+ \rightarrow Fe^{3+} + \tfrac{1}{2}H_2O \qquad (5.36)$$

$$FeS_2 + 14Fe^{3+} + 8H_2O \rightarrow 15Fe^{2+} + 2SO_4^{2-} + 16H^+ \qquad (5.37)$$

The reactions for Zn and Cu ores can be written in a simplified form as:

Table 5.15   Common ore minerals of non-ferrous metals (Peters, 1978; Rose *et al.*, 1979)

Metal	Ore minerals	Associated heavy metals
Ag	Ag_2S, PbS	Au, Cu, Sb, Zn, Pb, Se, Te
As	FeAsS, AsS	As, Au, Ag, Sb, Hg, U, Bi,
	Cu ores	Mo, Sn, Cu
Au	Native Au, $AuTe_2$	Te, Ag, As, Sb, Hg, Se
	$(Au,Ag)Te_2$	
Ba	$BaSO_4$	Pb, Zn
Bi	Pb ores	Sb, As
Cd	ZnS	Zn, Pb, Cu
Cr	$FeCr_2O_4$	Ni, Co
Cu	$CuFeS_2$, Cu_5FeS_4	Zn, Cd, Pb, As, Se, Sb, Ni, Pt, Mo, Au, Te
	Cu_2S, Cu_3AsS_4	
	CuS, Native Cu	
Hg	HgS, Native Hg	Sb, Se, Te, Ag, Zn, Pb
	Zn ores	
Mn	MnO_2	Various (eg Fe, Co, Ni, Zn, Pb)
Mo	MoS_2	Cu, Re, W, Sn
Ni	$(Ni,Fe)_9S_8$, NiAs	Co, Cr, As, Pt, Se, Te
	$(Co,Ni)_3S_4$,	
Pb	PbS	Ag, Zn, Cu, Cd, Sb, Tl, Se, Te
Pt	Native Pt, $PtAs_2$	Ni, Cu, Cr
Sb	Sb_2S_3, Ag_3SbS_3	Ag, Au, Hg, As
Se	Cu ores	As, Sb, Cu, Ag, Au
Sn	SnO_2,	
	$Cu_2(Fe, Zn)SnS_4$	Nb, Ta, W, Rb
U	U_3O_8	V, As, Mo, Se, Pb, Cu, Co, Ag
V	V_2O_5, VS_4	U
W	WO_3, $CaWO_4$	Mo, Sn, Nb
Zn	ZnS	Cd, Cu, Pb, As, Se, Sb, Ag, Au, In

$$\text{Sphalerite: } ZnS + 2O_2 \rightarrow Zn^{2+} + SO_4^{2-} \qquad (5.38)$$

$$\text{Chalcopyrite: } CuFeS_2 + 4O_2 \rightarrow Cu^{2+} + Fe^{2+} + SO_4^{2-} \qquad (5.39)$$

Frequently, bacteria such as *Thiobacillus thiooxidans* help to catalyse the oxidation reactions in tailings deposits and in weathering orebodies.

When the sulphide ore minerals are smelted to obtain the metals, it is necessary to roast the ores in air in order the convert the sulphides to oxides which are then reduced to the metal. This process gives rise to large amounts of SO_2 and metal smelters are among the major sources of this important atmospheric pollutant. Although smelter fumes can now be scrubbed to remove the SO_2 this is by no means the practice everywhere and much severe environmental pollution has occurred and still occurs from this source. In the past, the acidic fumes together with metal aerosol emissions exacerbated the environmental impact of the metals for many kilometres from the smelter source.

Another feature shown in Table 5.15 is that most of the major ore minerals often have several other metals associated with them. Many of these metals will have contaminated the environment in the vicinity of mines and smelters, although nowadays it is more likely that the minor constituents will also be refined and marketed as well as the major metal constituent of the ores. Old mine waste deposits have been reworked in some countries, such as the UK, where the mineral separations used in Victorian times were inefficient and some minerals in the ore deposit were considered worthless at that time (gangue minerals). For example, hydrothermal Pb–Zn deposits in Carboniferous limestone rocks in Derbyshire, England, contained large amounts of fluorspar (CaF_2) and barite ($BaSO_4$) which were dumped at the mines. With the increased demand for fluorine for use in potable water (NaF), PTFE non-stick coatings, and chlorofluorocarbon compounds in refrigerators and aerosol propellants (see p. 112) it became profitable to rework old mine waste tips. In the process, economically worthwhile amounts of barite and PbS were also extracted; the former was utilized in deep drilling as a high density medium (e.g. 4.5 g/cm), and the latter was ultimately smelted for Pb metal production. A consequence of the reuse of mined material anywhere in the world is the production of large quantities of tailings which need to be disposed of in an environmentally appropriate manner. Modern mineral separation methods involve the use of large volumes of water but much of this is normally recycled within the process, although smaller volumes of effluents containing metals, frothing agents and other chemicals (including cyanides in gold extraction) do need disposal eventually.

Agricultural materials.   Agriculture constitutes one of the very important non-point sources (NPS) of metal pollutants. The main sources are:

- impurities in fertilizers: Cd, Cr, Mo, Pb, U, V, Zn (e.g. Cd and U in phosphatic fertilizers);
- pesticides: Cu, As, Hg, Pb, Mn, Zn (e.g. Cu, Zn and Mn-based fungicides, Hg seed dressings, historical Pb–As orchard sprays);
- desiccants: As for cotton;
- wood preservatives: As, Cu;
- wastes from intensive pig and poulty production: Cu, As;
- composts and manures: Cd, Cu, Ni, Pb, Zn, As;
- sewage sludge: especially Cd, Ni, Cu, Pb, Zn (but many other elements);
- corrosion of metal objects (e.g. galvanized metal roofs and wire fences: Zn, Cd).

Fossil fuel combustion.   A wide range of heavy metals is found in fossil fuels which are either emitted into the environment as particles during combustion (see section 6.1), or accumulate in ash which may itself be transported and contaminate soils or waters, or may be leached *in situ*. Some of the metals arising as pollutants from fossil fuel combustion are Pb,

Cd, Zn, As, Sb, Se, Ba, Cu, Mn and V. The combustion of petrol (gasoline) containing Pb additives gives rise to large amounts of Pb particulates, mainly PbBrCl. Lead-containing particles in the exhausts of petrol vehicles are 0.01–0.1 μm in diameter, but these primary particles can cluster to form larger particles (0.3–1 μm). Diesel smoke normally contains these larger size particles. The lead-containing particles in motor vehicle exhausts tend to be larger in rural areas and near motorways than in urban areas (Fergusson, 1990).

Coal combustion gives rise to a wide range of metals, including U in some coal which accounts for coal-burning power stations being responsible for the emission of significant amounts of radioactive pollutants. Coal ash can contain relatively high concentrations of soluble compounds, including oxides of B, As, Se which can be leached and cause toxicity in sensitive crops and aquatic organisms. Coal ash is an important source of Cr (<172 μg/g). Crude oil can contain relatively large amounts of V which is emitted during the combustion of some oil products and accumulates in the ash of oil-fired boilers.

Steinnes (1987) reported concentrations of Pb, Cd, Zn, As, Sb and Se to be around ten times higher in moss, soil humus and topsoils along the southern coast of Norway than in the centre of the country. This was ascribed to the deposition of aerosols arising largely from fossil fuel combustion in north-western Europe.

Metallurgical industries. Many heavy metals are used in specialist alloys and steels: V, Mn, Pb, W, Mo, Cr, Co, Ni, Cu, Zn, Sn, Si, Ti, Te, Ir, Ge, Tl, Sb, In, Cd, Be, Bi, Li, As, Ag, Sb, Pr, Os, Nb, Nd and Gd. Hence both the manufacture and disposal, or recycling, of these alloys in scrap metal can lead to environmental pollution of a wide range of metals. Steel manufacture usually involves a lot of recycling of scrap and so steel works are often discrete point sources of atmospheric aerosols of metals.

Non-ferrous metal production causes marked environmental pollution not only of the metals being manufactured, but also of other minor associated metals (see Table 5.13) such as As, Cd, Cr, Cu, Co, Ni, Pb, Sb, Tl, Te, U, V, Zn and Se. Nowadays, many of these metals are extracted from the ores, refined and sold, but in the past this was not usually the case. For example, Cu smelters have a long history of causing As pollution in the surrounding countryside. Estimates of 1.5 kg As emitted per tonne of Cu produced are typical, with values of up to 16.8 kg As/t Cu from time to time at a smelter in Washington State, USA (O'Neill, 1990). In Austria, cancerous conditions have been reported in livestock grazing near to Cu smelters. Zinc ores often contain relatively high concentrations of Cd ($<5\%$) and so Zn production (and, to a lesser extent, Pb, and Cu smelting) can give rise to significant environmental pollution with Cd. Pacyna (1987) estimated that primary non-ferrous metal production gave rise to atmospheric emissions of 1630 t Cd/yr. Anomalously high Cd concentrations were found in soils and vegetation up to 40 km downwind from historic smelting activities in the Lower Swansea Valley in South Wales and 15 km from the Avonmouth Pb–Zn Smelter near Bristol, in the UK.

Electronics.   A large number of trace elements, including the heavy metals, are used in the manufacture of semi-conductors and other electrical components. These include Cu, Zn, Au, Ag, Pb, Sn, Y, W, Cr, Se, Sm, Ir, In, Ga, Ge, Re, Sn, Tb, Co, Mo, Hg, Sb, As and Gd. Environmental pollution can occur from the manufacture of the components and their disposal in waste. There is now a growing industry concerned with the recovery of valuable metals from decommissioned items of complex electrical equipment, such as computers. However, it must be remembered that old electronic equipment will include capacitors and transformers containing PCBs, which are highly toxic and persistent organic environmental pollutants (see section 6.4).

Other sources.   Other significant sources of heavy metal pollution in manufacture (sometimes in use) and disposal include:

Batteries – Pb, Sb, Zn, Cd, Ni, Hg, Pm
Pigments and paints – Pb, Cr, As, Sb, Se, Mo, Cd, Ba, Zn, Co, I, Ti
Catalysts – Pt, Sm, Sb, Ru, Co, Rh, Re, Pd, Os, Ni, Mo, I
Polymer stabilizers – Cd, Zn, Sn, Pb (from incineration of plastics)
Printing and graphics – Se (xerox process), Pb, Cd, Zn, Cr, Ba
Medical uses: dental alloy – Ag, Sn, Hg, Cu and Zn
              drugs/medicinal preparations – As, Bi, Sb, Se, Ba, Ta, Li, Pt
Additives in fuels and lubricants – Se, Te, Pb, Mo, Li

Waste disposal.   Many metals, especially Cd, Cu, Pb, Sn and Zn, are dispersed into the environment in leachates from landfills, which pollute soils and groundwaters, and in fumes from incinerators (see chapter 7). Sewage sludge contains many metals, including Zn, Cu, Pb, Cr, As and Mo, but the greatest cause for concern is currently considered to be Cd. Although present in many sludges in quite low concentrations (<10 μg/g), Cd is relatively easily taken up by food crops, especially leafy vegetables, and enters the human diet. Under the European Community Directive 86/278, the maximum permissible concentration of Cd in sludged soils used for food production is 3 μg/g (see chapter 7).

5.3.4   Environmental media affected

Heavy metals originating from the sources listed above generally have their most significant effects on the following environmental media:

QMW LIBRARY
(MILE END)

Mining	air (fumes, ore dusts and fine tailings)
	waters (effluents and tailings)
	soils (waste tips, deposited dusts and fumes)
Agriculture	air (fungicide droplets)
	waters (fungicide spillages and wash-off, livestock, slurry, Cu, As, spillage and seepage)
	soils (fertilizers, wastes, manures and composts)
Fossil fuel combustion	air (aerosol particles from combustion)
	water (ash-pollutants leached into water courses)
	soils (deposited aerosol particles, ash disposal and leaching)
Metallurgical industries	air (aerosol particles from furnaces, dusts from resuspension of deposited larger particles)
	waters (effluents, wash-off of particles)
	soil (deposited aerosols and larger particles, metal-rich sewage disposal, waste dumps)
Electronics	air (aerosols from manufacturing processes)
	waters (effluents and corrosion of decommissioned electrical components)
	soils (wastes and corrosion of decommissioned electrical components)
Chemical industries	air (volatilization of electrodes and catalysts, explosions)
	waters (effluents, spillages)
	soils (wastes)
Pigments and paints	air (droplets of sprayed paint, particles of weathered paints)
	waters (effluents, anti-fouling paints, weathering of pigments and paints)
	soils (wastes, spillages, weathering of pigments and paints)
Waste disposal	air (aerosols from incineration of metal containing wastes)
	waters (leachates from landfill, runoff, corrosion of waste dumped in wet pits)
	soils (disposal/utilization of waste, e.g. sewage, composts from wastes, fallout of aerosols from incinerators, ash from bonfires)

5.3.5 Heavy metal behaviour in the environment

On reaching the environment after emission, or being released, the metal pollutants behave in the following ways.

Atmospheric aerosol particles. These remain suspended for varying lengths of time determined by the size of the particle, the windspeed, relative humidity and precipitation (see section 6.1). Aerosol particles range in diameter from 5 nm to 20 μm but most are in the size range 0.1–10 μm. Particles >10 μm tend to settle out under gravity relatively rapidly but those <10 μm remain in the atmosphere for 10–30 days being removed by washout, settlement, impaction and, in the case of very small particles (<0.3 μm) by diffusive deposition. Under some circumstances, such as high humidity, smaller particles may sometimes cluster and form larger particles which are deposited more rapidly. In the 10–30 day period during which aerosol particles may remain suspended in the atmosphere they can be transported thousands of kilometres, depending on the circulation of air masses. It is for this reason that elevated concentrations of Pb, Cd, Zn, As and Se are found in the soils of southern Norway due to the transport of particles from the industrialized areas of Europe.

While suspended in the air, these metal aerosol particles may be inhaled by humans and animals and subsequently absorbed into the bloodstream through the alveoli of the lungs. Particles falling onto foliage may also enter plant tissues by absorption through the cuticle but this depends on the presence of moisture and its pH, the type of plant and other parameters. For example, children are at severe risk from the inhalation of aerosol sized particles of lead compounds in the exhaust emissions of motor cars. These particles also settle onto crop foliage and may be consumed with the plant (e.g. on lettuce leaves). Most particles reach the soil eventually and may be ingested with incompletely washed vegetables, by children eating soil intentionally (pica), or accidentally from unwashed adult or children's hands after gardening or playing with contaminated soil.

On being deposited onto the soil, the metal compounds in the aerosol particles react with the soil constituents and become incorporated into the soil system. Metal particles absorbed by plants also reach the soil through the mineralization of plant litter.

Aqueous and marine environments. Aerosol particles deposited into water, either directly or washed off surfaces into water courses, either react with the constituents of the water or settle to the bottom where they react with the sediments. The solubility of metal ions in solution will depend on the concentrations of anions and chelating ligands, present in the water, its pH and redox status, and the presence of adsorbent sediments. Several metal ions are adsorbed and coprecipitated with hydrous oxides of Fe, Mn and Al, in both sediments and soils, for example Fe oxides coprecipitate V, Mn, Ni, Cu, Zn, Mo; Mn oxides coprecipitate Fe, Co, Ni, Zn, Pb.

Calcium carbonate, either originating from limestone rock fragments in soils and sediments, or by precipitation from soil water in the soils of semi-arid and arid regions, also adsorbs a range of metals, including V, Mn, Fe, Co and Cd. Clay minerals in soils and sediments are responsible for the adsorption and coprecipitation of V, Ni, Co, Cr, Zn, Cu, Pb, Ti, Mn and Fe.

Metal ions in solution can be absorbed into aquatic plants and animals and can cause toxicity if the concentration is sufficiently high. This factor is exploited in the use of $CuSO_4 \cdot 5H_2O$ to control algal blooms in lakes and reservoirs.

A survey of drinking water in the UK showed that first draw drinking water in 24 towns had Pb>50 µg/l. The worst situation was in Scotland where 34.4% of homes had Pb concentrations in water above the limit (Fergusson, 1990). This is due to the use of Pb pipes and solder and to the water being relatively acidic. This is clearly shown in the data from Packham (1990) who reported that in hard and soft waters from a total of 235 towns in the UK, the mean Cu in flushed water was 24 µg/l for towns with soft water and 14 µg/l for towns with hard water. Copper intake from drinking water can be around 1.4 mg/day from soft water and 0.05 mg/day from hard water (Davies and Bennett, 1983). The main differences in constituents between hard and soft waters are that hard waters contain high concentrations of Ca and Mg, (< 500 mg/l as $CaCO_3$) whereas soft

Table 5.16 Guideline and maximum acceptable concentrations for metal and other inorganic pollutants in drinking waters (WHO, 1984; Murley, 1992; Canada Council of Ministers of the Environment, 1991; Manahan, 1991) (µg/l) (Reproduced with permission of the Minister of Supply and Services Canada, 1993.)

	EC/WHO	Canada*	USA*
As	50(W)	25	50
B	1000	5000	1000
Ba	100	1000	—
Fe	50	<300	50
Mn	20	<50	50
Cd	5(W)	5	10
Cr	50(W)	50	50
CN	100(W)	200	—
Pb	50(W)	10	5
Hg	1(W)	1	—
Cu	<3000 (100 at works)	<1000	1000
U	—	100	—
Zn	<5000 (100 at works)	<5000	5000
F	1500 (8–10°C) 700 (25–30°C)	1500	800–1700 (depending on temperature)

EC = 80/778/EEC Quality of water for human consumption
(W) = WHO 1984 guideline values
* Maximum acceptable values

waters contain higher concentrations of Al, Mn and Pb but where soft waters are also acid, concentrations of copper are likely to be much higher. The pH of domestic water from public supplies should be between 6.5 and 8.5. The metal content of water in houses is dependent upon the following factors (adapted from Mattson (1990)):

- original metal content, pH, alkalinity and oxygen contents, total hardness and temperature of the water;
- duration of time the water remains in the pipe (still-stand);
- length of flushing time after still-stand;
- materials, age, internal diameter and total length of the pipes in the house, and the type of solder.

Heavy metal ions in soils.   Heavy metal pollution can affect all environments but its effects are most long lasting in soils due to the relatively strong adsorption of many metals onto the humic and clay colloids in soils. The duration of contamination may be for hundreds and thousands of years in many cases (e.g. first half lives: Cd, 15–1100 years, Cu 310–1500 years and Pb 740–5900 years depending on the soil type and their physico-chemical parameters). Unlike organic pollutants, which will ultimately be decomposed, metals will remain as metal atoms although their speciation may change with time as the organic molecules binding them decompose or soil conditions change.

The extent to which metal ions are adsorbed by cation exchange and non-specific adsorption depends on the properties of the metal concerned (valency, radius, degree of hydration and coordination with oxygen), pH, redox conditions, the nature of the adsorbent (permanent and pH-dependent charge, complex-forming ligands), the concentrations and properties of other metals present, and the presence of soluble ligands in the surrounding fluids (see section 2.6.2).

The selectivity of clay mineral and hydrous oxide adsorbents in soils and sediments for divalent metals generally follows the order Pb > Cu > Zn > Ni > Cd, but some differences occur between minerals and with varying pH conditions. The selectivity order for peat has been shown to be Pb > Cu > Cd = Zn > Ca. However, in general, Pb and Cu tend to be adsorbed most strongly and Zn and Cd are usually held more weakly, which implies that these latter metals are likely to be more labile and bioavailable (see Table 5.17).

It is usually found that the adsorption of metal ions onto soil solids is described by either the Langmuir or the Freundlich adsorption isotherm equations. The Langmuir equation is:

$$\frac{M}{x/m} = \frac{1}{Kb} + \frac{M}{b}$$

where M is the activity of the ion, x/m is the amount of M adsorbed per unit of adsorbent, K is a constant related to the bonding energy and b is the maximum amount of ions that will be adsorbed by a given adsorbate.

The Freundlich equation is:

$$\log x = k + n \log c$$

where x is the amount adsorbed per unit of adsorbent at concentration c of adsorbate and k and n are constants. These isotherms do not provide any information about the adsorption mechanisms involved and both assume a uniform distribution of adsorption sites on the adsorbent and absence of any reactions between adsorbed ions.

In general, nearly all metals (except Mo) are most soluble and bioavailable at low pHs and therefore toxicity problems are likely to be more severe in acid environments. In the case of pollution by particles of sulphide ore minerals, the weathering of the sulphide exacerbates the problem by increasing the acidity of the soil. In agricultural soils this situation can be mitigated to a considerable extent by liming.

Methylation of heavy metals in the environment.   Arsenic, Hg, Co, Se, Te, Pb and Tl can be methylated in the environment through the action of enzymes secreted by microorganisms (biomethylation) and also by abiotic chemical reactions. There is a possibility that Cd, In, Sb and Bi can also be methylated. The bacteria associated with methylation of these elements are found in the bottom sediments of rivers, lakes and the coastal waters of the sea, soils and the digestive tracts of animals (including humans). This methylation radically affects the behaviour of the elements in the components of the environment, their bioavailability and their toxicology. For example, monomethyl mercury (CH_3Hg^+), the most toxic form of Hg, is lipophyllic and therefore accumulates in body fats and is the only heavy metal to show bioaccumulation along the food chain (Fergusson, 1990).

The uptake of heavy metals by plants.   Transfer coefficients (concentration of metal in aerial portion of plant relative to total concentration in the soil) are a convenient way of quantifying the relative differences in bioavailability of metals to plants. Kloke *et al.* (1984) gave generalized transfer coefficients for soils and plants (Table 5.17); however, soil pH, soil organic matter content and plant genotype can have marked effects on metal uptake. The transfer coefficients are based on root uptake of metals but it should be realized that plants can accumulate relatively large amounts of metals by foliar absorption of atmospheric deposits on plant leaves.

From Table 5.17 it can be seen that Cd, Tl and Zn have the highest transfer coefficients which is a reflection of their relatively poor sorption in

Table 5.17 Transfer coefficients of heavy metals in the soil–plant system (from Kloke *et al.* (1984))

Element	Transfer coefficient	Element	Transfer coefficient
As	0.01–0.1	Ni	0.1–1
Be	0.01–0.1	Pb	0.01–0.1
Cd	1–10	Se	0.1–10
Co	0.01–0.1	Sn	0.01–0.1
Cr	0.01–0.1	Tl	1–10
Cu	0.01–0.1	Zn	1–10
Hg	0.1–1		

the soil. In contrast, metals such as Cu, Co, Cr and Pb have low coefficients because they are usually strongly bound to the soil colloids.

5.3.6   Toxic effects of heavy metals

The sensitivity of organisms to metal toxicity varies widely with species of plants and animals and genotypes within species (e.g. cultivars of crops) and many factors can modify the response to the toxic dose of metals. Some individuals are genetically adapted to tolerating anomalously high concentrations of certain metals. Homeostatic mechanisms in animals frequently involve special proteins, metallothioneins, which bind with the metals (such as Cd) and render them relatively inactive. Plants have similar compounds called phytochelatins.

It is therefore difficult to generalize about toxicity. However, an indication of relative toxicity of metals to mammals is provided by the LD50 values for a wide range of heavy metals in Table 5.20.

Phytotoxicity.   The most toxic metals for both higher plants and several microorganisms are Hg, Cu, Ni, Pb, Co, Cd, and possibly Ag, Be and Sn (Kabata-Pendias and Pendias, 1984). Although the occurrence of toxicity will depend on soil factors, such as pH, the plant genotype and the conditions under which it is growing (pot in greenhouse, or under field conditions), a general indication of the toxic levels of some metals is given in Table 5.18.

The data in Table 5.20 provide an indication of the relative toxicity of different elements but this will be affected by considerable variation in toleration of individuals and, in the case of diets, in the composition of the diets. The doses injected into rats and other experimental mammals probably provide a more accurate comparison of toxicity. From this data for injected doses, U and ^{239}Pu are jointly the most toxic elements, followed by Cd and Se and then Hg. From the human diet data, As, Cu

Table 5.18 Normal and phytotoxic metal concentrations generally found in plant leaves (Alloway (1990), based largely on Bowen (1979))

	Concentration in leaves (μg/g)	
Element	Normal range	Toxicity
Ag	0.01–0.8	1–4
As (III)	0.02–7	5–20
Cd	0.1–2.4	5–30
Cu	5–20	20–100
Cr	0.03–14	5–30
Hg	0.005–0.17	1–3
Ni	0.02–5	10–100
Pb	5–10	30–300
Sb	0.0001–2	1–2
V	0.001–1.5	5–10
Zn	1–400	100–400

Table 5.19 Biochemical effects of excessive concentrations of heavy metals in plants (from Kabata-Pendias and Pendias (1984) and reprinted from Fergusson, *The Heavy Elements: Chemistry, Environmental Impact and Health Effects*, © 1990, p. 40, with kind permission from Pergamon Press Ltd, Headington Hill Hall, Oxford OX3 0BW, UK)

Elements	Biochemical process affected
Ag, Au, Cd, Cu, Hg, Pb, F, I, U	Changes in the permeability of cell membranes
Hg	Inhibition of protein synthesis
Ag, Hg, Pb, Cd, Tl, As(III)	Bonding to sulphydryl groups
As, Sb, Se, Te, W, F	Competition for sites with essential metabolites
Most heavy metals, Al, Be, Y, Zr, lanthanides	Affinity for phosphate groups, and ADP, ATP groups
Cs, Li, Rb, Se, Sr	Replacement of essential atoms
Arsenate, selenate, tellurate	Occupation of sites for essential groups, e.g. PO_4^{3-}, tungstate, bromate, fluorate
Tl, Pb and Cd	Inhibition of enzymes
Cd, Pb	Respiration
Cd, Pb, Hg, Tl, Zn	Photosynthesis
Cd, Pb, Hg, Tl, As	Transpiration
Cd, Co, Cr, F, Hg, Mn, Ni, Se, Zn	Chlorosis
Al, Cu, Fe, Pb, Rb	Dark green leaves

and Hg are the most toxic but homeostatic mechanisms will often affect the extent of toxicity.

Examples of critical (trigger) concentrations of heavy metals used in different countries are given in Tables 5.21–5.23.

Table 5.21 shows the critical concentrations used in The Netherlands for

Table 5.20 Relative mammalian toxicity of elements in injected doses and diets (from Bowen (1979) with permission)

Element	Acute lethal doses (LD_{50}) injected into mammals* (mg/kg bodyweight)	Dose in human diet (mg/day)	
		Toxic	Lethal
Ag	5–60	60	1.3k–6.2k
As	6	5–50	50–340
Au	10	—	—
Ba	13	200	3.7k
Be	4.4	—	—
Cd	1.3	3–330	1.5k–9k
Co	50	500	—
Cr	90	200	3k–8k
Cs	1200	—	—
Cu	—	—	175–250
Ga	20	—	—
Ge	500	—	—
Hg	1.5	0.4	150–300
Mn	18	—	—
Mo	140	—	—
Nd	125	—	—
Ni	110–220	—	—
Pb	70	1	10k
Pt	23	—	—
(^{239}Pu)	1	—	—
Rh	100	—	—
Sb	25	100	—
Se	1.3	5	—
Sn	35	2000	—
Te	25	—	2k
Th	18	—	—
Tl	15	600	—
U	1	—	—
V	—	18	—
Zn	—	150–600	6k

* Injected into the peritoneum to avoid absorption through the digestive tract. Chemical form of the element will affect its toxicity.

contaminated soils and waters. Until recently a system was used for soils which involved three indicative values: A, the 'normal' reference value; B, the test value to determine the need for further investigations; and C, the intervention value above which the soil definitely needs cleaning-up (Moen *et al.*, 1986). This system has been superseded by an effect-oriented scheme of 'Environmental Quality Standards for Soil and Water' (Netherlands Directorate General for Environmental Protection, 1991). These standards are based on ecological function and comprise target values (TV) for soils which represent the final environmental quality goals for the Netherlands. Both target and limit values for waters are given because it is intended that the target value is reached by progressivedly lowering the limit values. This is easier for waters than for soils owing to the long residence time of most

Table 5.21 Guide values and quality standards used in The Netherlands for assessing soil and water contamination by heavy metals (Netherlands Ministry of Housing, Physical Planning and Environment, 1991)

| Metals | Soils (µg/g) | | | | Waters (µg/l) | | | | |
| | | | | | Surface waters | | | | Groundwater |
	*A	B	C	STV	TotTV	TotLV	DisTV	DisLV	GWDisTV
As	20	30	50	29	5	10	4	8.6	10
Ba	200	400	2000	200	—	—	—	—	50
Cd	1	5	20	0.8	0.05	0.2	0.003	0.005	1.5
Co	20	50	300	10	—	—	—	—	10
Cr	100	250	800	100	5	20	0.5	2	1
Cu	50	100	500	36	3	3	1	1.3	15
Hg	0.5	2	10	0.3	0.02	0.03	0.003	0.005	0.05
Mo	10	40	200	10	—	—	—	—	5
Ni	50	100	500	35	9	10	7	7	15
Pb	50	150	600	85	4	25	0.2	1.3	15
Sn	20	50	300	20	—	—	—	—	10
Zn	200	500	3000	140	9	10	2	2	150

Soil values: A = reference value, B = test requirements, C intervention value, from 1986 Scheme; STV = target value for soils; surface waters TotTV = total content target value; TotLV = total content limit value; DisTV = dissolved content target value; DisLV = dissolved content limit value; GWDisTV = groundwater dissolved content target value (from 1991 Environmental Quality Standards for Soils and Waters). Target values for soils are based on 'standard soil' (10% organic matter and 25% clay)

pollutants in soils. The intervention values (C) of the 1986 scheme still stand. In some cases the target values based on risk assessment are lower than the background levels (A values) given in the 1986 scheme. This is due in part to the wide differences in sensitivity to the contaminants between organisms within ecosystems.

The critical values for soils given in Table 5.22 for the UK by the Department of the Environment Interdepartmental Committee for the Reclamation of Contaminated Land List of Trigger Concentrations for Contaminants (DOE, 1987) are more pragmatic and based mainly on the risk to human health.

The new Canadian National Classification System for contaminated soils shown in Table 5.23 is intended for use in the evaluation of contaminated sites. The values in the tables enable sites to be classified as high, medium or low risk according to their impact (current or potential) on human health and ecosystems. It is a screening system and is not intended to be a quantitative risk assessment for individual sites.

5.3.7 Analytical methods

The most commonly used analytical methods for determining the concentrations of heavy metals in extracts and acid digest are atomic absorption

Table 5.22 UK Department of the Environment ICRCL trigger concentrations for environmental metal contaminants (total concentrations except where indicated (UK Department of the Environment, 1987)

Contaminant	Proposed uses	Threshold trigger concentration (μg/g)
Contaminants which may pose hazards to human health		
As	Gardens, allotments	10
	Parks, playing fields, open space	40
Cd	Gardens, allotments	3
	Parks, playing fields, open space	15
Cr (hexavalent*)	Gardens, allotments	25
	Parks, playing fields, open space	—
Cr (total)	Gardens, allotments	600
	Parks, playing fields, open space	1000
Pb	Gardens, allotments	500
	Parks, playing fields, open space	2000
Hg	Gardens, allotments	1
	Parks, playing fields, open space	20
Se	Gardens, allotments	3
	Parks, playing fields, open space	6
Phytotoxic contaminants not normally hazardous to health		
B (water soluble)	Any uses where plants grown	3
Cu (total)	Any uses where plants grown	130
(extractable**)		50
Ni (total)	Any uses where plants grown	70
(extractable)		20
Zn (total)	Any uses where plants grown	300
(extractable)		130

* Hexavalent Cr extracted by 0.1M HCl adjusted to pH at 37.5°C
** Extracted in 0.05M EDTA

spectrophotometry (AAS) or flame AAS for most jobs (0.1–10 μg/ml) but electrothermal atomization AAS (also called graphite furnace AAS) is more useful for low concentrations (in some cases down to ng/ml). Hydride generation equipment can be linked to AAS to allow As, Hg, Sb and certain other elements to be determined. Multi-element analysis (< 22 elements) can be carried out by inductively coupled plasma – atomic emissions spectrometry (ICP-AES) and by X-ray fluorimetry (see chapter 4).

5.3.8 Examples of specific heavy metals

Arsenic. A toxic, non-essential element that has been used as a pigment, pesticide, wood preservative and a livestock growth promoter (pigs and poultry, phenylarsonic acid) and has caused environmental pollution in these roles. It is also present in many sulphide ores of metals and is therefore emitted from metal smelters as an atmospheric pollutant (40% of

Table 5.23 Selected values from the new Canadian interim environmental quality criteria for soil ($\mu g/g$) (Canadian Council of Ministers of the Environment, Winnipeg, 1991)

Metal	Background	Agricultural	Residential	Industrial
As	5	20	30	50
Ba	200	750	500	2000
Be	4	4	4	8
Cd	0.5	3	5	20
Cr^{6+}	2.5	8	8	*
Co	10	40	50	300
Cu	30	150	100	500
CN (free)	0.25	0.5	10	100
CN (total)	2.5	5	50	500
Pb	25	375	500	1000
Hg	0.1	0.8	2	10
Mo	2	5	10	40
Ni	20	150	100	500
Se	1	2	3	10
Ag	2	20	20	40
Sn	5	5	50	300
Zn	60	600	500	1500

* Criteria not recommended

the anthropogenic emissions) (O'Neill, 1990). Coals also contain significant amounts of As and its combustion accounts for 20% of atmospheric emission. Coal ashes are also a significant source of As which can be leached out into waters or the soil. The toxicological importance of As is partly due to its chemical similarity with P which means that As can disrupt metabolic pathways involving P. Both acute and chronic toxicity are recognized and the continual inhalation of airborne forms of As is known to be carcinogenic. Respiratory cancers have occurred in occupationally exposed workers.

Cadmium.   A highly toxic non-essential metal which accumulates in the kidneys of mammals and can cause kidney dysfunction. In humans, kidney damage diagnosed by the presence of microglobulin proteins is the main toxic effect resulting from chronic exposure to the metal. High concentrations of inhaled Cd aerosols can cause emphysema and related acute lung conditions. Cadmium becomes very volatile above 400°C and hence is likely to be dispersed as an aerosol when mixtures of metals containing Cd are heated or cast. Cadmium is a 'modern' metal, having been used increasingly in corrosion prevention, polymer stabilization, electronics and pigment applications within the last 30 years. It tends to be less strongly adsorbed than many other divalent metals and is therefore more labile in soils and sediments and more bioavailable. There is more danger from this metal moving through the human food chain from contaminated soils than most other metals. Sewage sludge-amended soils can contain sufficiently high concentrations of Cd to cause elevated concentrations of Cd in food crops and there is a European Community Directive limiting the maximum Cd content of sludged soils to 3 μg/g. Although sewage sludge applications are considered a major source of Cd in the soils receiving sludges, the most important sources overall are phosphatic fertilizers and industrial emissions.

A serious case of Cd poisoning occurred in the Jintsu Valley in the Toyama Prefecture in Japan, where Pb–Zn mining had caused widespread Zn and Cd contamination of the alluvial soils, most of which were used for paddy rice production. The farmers in the valley live mainly on rice grown in the contaminated paddies and also relied on the metal-polluted river for their drinking water. After the Second World War, it was found that more than 200 elderly women who had all had several children had developed kidney damage and skeletal deformities. The condition was known as 'itai-itai' disease which literally means 'ouch-ouch' due to the pain caused by the deformed bones. The Cd toxicity was exacerbated by a low protein and vitamin D diet and the birth of several children. The rice which they consumed contained ten times more Cd than local controls and the contaminated water was an additional intake. It was estimated that the people in the valley had a Cd intake of around 600 μg Cd/day which is around ten times greater than the maximum tolerable intake of 60–70 μg/day. A survey of paddy soils in the whole of Japan revealed that 9.5% of the area was significantly contaminated with Cd, with a further 3.2% of upland rice-growing soils and 7.5% of orchard soils. The source of the Cd in these soils is probably phosphatic fertilizers and industrial/mining pollution.

Copper.   A micronutrient which can be deficient in some soils causing severe loss of yield in several crops, especially cereals. Toxicity problems can occur in crops in polluted soils and in livestock grazing herbage growing on polluted soils. Sheep are the agricultural livestock most sensitive to Cu toxicity, but they (and cattle) are also prone to deficiency disorders. Herbage with < 5 μg/g Cu can lead to Cu deficiency in both sheep and cattle but if the herbage contains > 10 μg/g Cu then toxicity is likely to occur in sheep. Copper pollution can arise from Cu mining and smelting, brass manufacture, electroplating and excessive use of Cu-based agrichemicals (e.g. Bordeaux Mixture). Copper sulphate is used widely as an algicide in ornamental ponds and even in water supply reservoirs which are affected by blooms of toxic blue-green algae. Copper is used widely in houses for piping water and although the concentrations of Cu in the drinking water is higher in soft water, this is not considered to be a hazard so long as the pH is within the normal limits (pH 6.5–8.5). More acid waters could create problems with excessive concentrations, but none have been reported with public water supplies.

Chromium.   A micronutrient, which is essential for carbohydrate metabolism in animals; pollution of soils occurs as a result of the dumping of chromate wastes, such as those from tanneries or electroplating and from sewage sludge disposal on land. However, unlike many other heavy metals, Cr can exist in a trivalent (chromite) and a hexavalent form (chromate) and the Cr(VI) form is more phytotoxic than the Cr(III) form. Soluble chromate concentrations of 0.5 μg/ml have been shown to be significant. Hence the redox conditions in the environment are very important, waterlogged soils with reducing soils will have the less toxic Cr(III). However, in many freely drained aerated soils the predominant form is also Cr(III) because soil organic matter results in the reduction of Cr(VI) to Cr(III). Soils developed on ultramafic rocks, such as serpentinites can contain very high concentrations of Cr of geochemical origin and cannot be considered polluted but either 'naturally' or 'geochemically' enriched.

Chromite (III) appears to be more toxic to fish than Cr(VI), especially salmon, but toxic concentrations for several species of fish range from 0.2–5 μg/ml. Municipal wastewater can contain concentrations of <0.7 μg/Cr ml, mainly in the Cr(VI) form, which are toxic to many species of marine animals, algae and microorganisms but reduction of Cr(VI) to Cr(III) usually occurs if there is organic matter present (Langard, 1980).

Chromium is carcinogenic, causing cancer of the respiratory organs in chromate workers chronically exposed to Cr-containing dusts (Langard, 1980).

Mercury.   A non-essential element. An important source of air pollution is the chlor-alkali process for the production of Cl_2 and NaOH from brine (p. 221) where 0.1 to 0.2 kg of Hg are lost into the environment with every 1000 kg of Cl_2 produced. Mercury is used as a catalyst in the production of some plastics and there are cases of severe environmental pollution resulting from the discharge of Hg-containing liquid wastes. In the Minamata Bay in Japan, $HgSO_4$ in effluents reaching the bay was first of all precipitated as insoluble HgS which later underwent biomethylation through the action of bacteria in the sediments to form CH_3Hg^+. This methylated form is very volatile and lipophyllic and accumulated in the food chain of fish resulting in the fish catches in the bay having high concentrations of Hg which were harmful to humans eating them. Mercury is a neurotoxin and has teratogenic effects; at least 78 people were severely affected by the methyl mercury poisoning, seven of them fatally. Another coastal area nearby, Niigata, was similarly affected. Other heavy metals can be methylated in the environment, but Hg appears to cause the most dangerous problems for human health (Fergusson, 1990). Other sources of Hg pollution in lakes and rivers is the use of Hg-containing slimicides in pulp paper mills.

Several cases of pollution from this source have been reported around the world, including Canada and Sweden.

Until recently, alkyl-Hg was used as an agricultural seed dressing to prevent fungal disease in germinating seeds. This resulted in significant amounts of the metal being added to highly productive, intensively farmed agricultural soils in technically advanced countries. Its use has been discontinued due to its toxicity to humans and wildlife. A serious case of mass poisoning (probably the most serious case of chemical poisoning through the diet) occurred in 1971–72 when a consignment of Hg-treated seed exported to Iraq for growing was mistakenly milled into flour and used for making bread. By the end of 1972, 6530 people had been admitted to hospital suffering from Hg poisoning and 459 of these patients died (Kazantis, 1980). Other modern sources of Hg are small batteries for use in cameras, hearing aids, etc. and these could easily be swallowed by small children.

Lead.   A non-essential element; it is a neurotoxin and a good example of a multimedia pollutant. The main sources of Pb pollution in the environment are petrol (air pollutant, but can also be water or soil pollutant from spillages), particulates in exhaust fumes from petrol combustion (air pollutants, inhaled by humans), particulates from petrol, fossil fuel combustion in soil (soil pollutant – taken up by plants and also ingested with plant food crops), paint flakes from old paint containing a percentage of Pb, Pb in some traditional ethnic cosmetics (e.g. surma – skin absorption), constituent of solders and varnishes used on interiors of food cans (food contaminant), Pb pipes for potable water (water pollutant), pesticide (historic use of Pb and As containing pesticide sprays in orchards), lead shot used in guns for game and clay pigeon shooting (soil pollutant, but also a food contaminant if inadvertently consumed with the game flesh) and, finally, Pb pollution from mining and smelting of the ore (usually PbS) – this includes acid mine drainage with soluble Pb (water pollutant), tailings from ore dressing (particulate water pollutant and soil pollutant, weather to release soluble Pb), and smelter fumes – Pb aerosols (air and soil pollutants).

On a comparative basis, Pb is neither as toxic as many other heavy metals (Tables 5.18 and 5.20) nor as bioavailable (Table 5.17), however, it is generally more ubiquitous in the environment and is a cumulative toxin in the mammalian body, so toxic concentrations can accumulate in the bone marrow, where red blood corpuscle formation (haematopoiesis) occurs. At least five stages in the formation of the haem part of haemoglobin are affected by Pb but the two enzymes most affected are δ-amino laevulinic dehydratase (ALAD) and ferrochetalase (Waldron, 1980). This inhibition of haem synthesis results in anaemia. Kidney damage also occurs as a result of exposure to Pb. Lead, like Hg, is a powerful neurotoxin and a range of pathological conditions are associated with acute Pb poisoning, most characteristic of which is cerebral oedema. However, the absorption of Pb in amounts which are not high enough to cause acute poisoning may induce behavioural abnormalities, including learning difficulties.

The critical concentrations for Pb in blood are the EC recommended level of 35 µg/dl and the UK threshold level for follow-up investigations of 25 µg/dl in at least half the population and less than 30 µg/dl in 90% of the population. The critical level in blood for occupational exposure is higher with values of 60, 80 and even 100µg/dl being used in different countries.

Zinc. A micronutrient which is the most serious deficiency problem in crops in the world as a whole, especially in tropical regions and where the soils have developed on sandstones and sandy drift. Humans can also be affected by deficiencies and in extreme cases short stature and delayed sexual maturity can be caused by it. In the context of pollution, Zn is mainly a cause of phytotoxicity and has a relatively low toxicity to animals and humans. Zinc pollution is often associated with mining and smelting and Cd is always present as a guest element in the ZnS ore. Mining causes pollution of air, water and soil with fine tailings particles which ultimately undergo oxidation to release Zn^{2+}. Zinc sulphide (sphalerite) ore often occurs together with PbS (galena), the main ore of Pb, and so Zn pollution is often associated with Pb and also Cu, in some cases, as well as traces of Cd. The geochemical association with Cd implies that impure Zn compounds may contain Cd, but also that Zn-Cd antagonism may mitigate some of the effects of Cd contamination.

A relatively ubiquitous source of Zn in the environment is galvanized steel which in the form of wire fencing gradually dissolves in rain and drops to the ground below, but in the case of roofing, the Zn ions usually drain away via gutters and drainpipes. Water collected from these roofs may not be suitable for drinking in areas of acid rain due to the high Zn and relatively high Cd concentrations which could arise.

**5.4
Other metals and
inorganic pollutants**

5.4.1 *Aluminium (atomic weight 27, comprises 8.2% of earth's crust)*

Aluminium is widely used as an alternative to steel for cladding structures, and as an alternative to copper in electricity and thermal conduction. Its oxide form (alumina) is widely used in industry as an abrasive. Aluminium hydroxide is also used as an antacid by people with digestive disorders. Aluminium is used extensively in water treatment and normally levels in the drinking water should be below 100 µg/l although the WHO guideline value for Al in drinking water is a compromise value of 200 µg/l (WHO, 1984). An accidental introduction of 20 tonnes of aluminium sulphate into the water supply of the Camelford District of Cornwall in July 1988 caused many reported health effects although allegations of impaired mental function have not been substantiated. Not only did this accident give rise to a very high Al concentration in the drinking water, but it also increased the acidity of the water which caused the solution of Cu from pipes and Pb from solders. In addition to the effects on people, thousands of fish were also killed when the erroneously contaminated water was discharged into a river.

Acid soil conditions exacerbated by acid precipitation (SO_x and NO_x) lead to increased concentrations of Al in stream and river waters (range 8–3500 µg/l), and these may be used for drinking water supplies.

Toxicity: Aluminium dust can cause lung disease (pulmonary fibrosis) which can lead eventually to emphysema. However, the majority of the population is exposed to aluminium mainly from food and drink, rather than from particulates in air, due to the use of Al in cooking pans, drink cans and as a constituent of foods and beverages. It is recognized that patients undergoing kidney dialysis could develop dementia as a result of the accumulation of Al from the water. Some investigations of the composition of the brains of people who had died from Alzheimer's disease (senile dementia) revealed high concentrations of aluminium. However, others have failed to find these accumulations of Al and so there is much uncertainty about a causal link between the metal and onset of this disease.

Exposure to aluminium could be large in some cases where people consume large doses of antacids (< 5000 mg Al/day) which is 42–250 times greater than the average daily intake from other sources. Drinking water obtained from rivers or lakes in acidified catchments may have considerably elevated Al concentrations. The tea plant is a natural accumulator of Al and so regular tea drinkers probably have an increased intake of Al (and also fluorine, which the plant also accumulates).

5.4.2   Beryllium (atomic number 4, atomic weight 9, concentration in the crust 2 µg/g)

Beryllium is the lightest of all the stable elements, it is resistant to corrosion, very hard and is used in various industrial applications, such as in alloys with Cu and Ni. Until 1949, it was used in phosphors in fluorescent lamps but this was discontinued owing to its hazardous nature. Coal contains small concentrations of Be (average 2.5 µg/g) and its combustion probably acts as the largest source of the metal in the environment. However, industries extracting or using the metal are the major sources of potentially hazardous concentrations.

Toxicity: The main hazard is in the inhalation of particles of the metal, its oxides or other compounds into the lungs. A chronic form of pneumo-coniosis called berylliosis appears many years after exposure (< 17 or even 25 years). Once the condition has commenced it is progressive, even in the absence of further exposure. An acute condition can develop within 72 hours of exposure to high concentrations of Be fumes, but cases are relatively rare. Although studies on the carcinogenicity of Be have been inconclusive, it is classified as a probable human carcinogen.

The main danger of exposure is in workers involved in the extraction, smelting or other uses of the metal, but residents living near to Be-using industries could also be at risk. The emission and transport of dusts and fumes from industrial plants using the metal therefore needs to be monitored and kept to an absolute minimum. However, the imposition of strict occupational hygiene standards has already reduced the risk of disease from the metal, but it remains a potential problem wherever it is manufactured or used (Harte *et al.*, 1991). Tepper (1980) has written a detailed review of the toxicology and environmental behaviour of Be.

5.4.3   *Fluorine (atomic number 9, atomic weight 18.9984, crustal abundance 0.0460%)*

Fluorine is a pale yellow gas (F_2) which is the most reactive of all elements and is used for the manufacture of various fluoride compounds such as UF_6, SF_6 and chlorofluorocarbons.

Various fluoride compounds cause environmental pollution problems, these include HF which can be generated during the manufacture of phosphatic fertilizers from the rock phosphate fluorapatite $Ca_5F(PO_4)_3$:

$$Ca_5F(PO_4)_3 + H_2SO_4 + 2H_2O = CaSO_4.2H_2O + HF + 3H_3PO_4$$

HF is very toxic and corrosive, it severely irritates the respiratory tract, eyes and other tissues and brief exposure to 1000 μg/ml may be fatal.

Silicon tetrafluoride SiF_4 is a pollutant from steelmaking plants using fluorspar CaF_2 as a flux:

$$2CaF_2 + 3SiO_2 = 2CaSiO_3 + SiF_4$$

Electrolytic Al smelting by the Hall–Heroult process uses cryolite (Na_3AlF_6) as a non-aqueous solvent for molten Al_2O_3 and consequently almost all Al smelters are significant sources of atmospheric F pollution. The method uses 70 kg (Na_3AlF_6) per tonne of Al metal produced (Fergusson, 1982).

Brick kilns are a common source of F compounds and they emit mixtures of HF, SiF_4 and H_2SiF_6 fumes and particulates (together with other pollutants including SO_2 and mercaptans). In the early 1980s two large brickworks situated in close proximity on the Oxford Clay in Bedfordshire, in the English Midlands, were emitting a combined total of 430 t F/year (and 43 000 t SO_2/yr). Average concentrations of F in the air around the works were in the range 1–8 μg/m, with maximum short period concentrations of < 35 μg/m (Owen, 1981 and S. Cray pers. comm.). Fluoride-containing particulates settle out onto pastures surrounding the brickworks and are ingested by grazing livestock causing a toxic condition called fluorosis, which affects the bones and teeth. Plants, including cereals and trees, can also be affected by F toxicity and a synergistic effect of combined SO_2 and F pollution is recognized in crops in areas affected by atmospheric pollution from brickworks. The Oxford Clay used for brickmaking in Bedfordshire contains a relatively high concentration of organic matter which is beneficial in brickmaking because it increases the calorific value of the clay, but it is also a source of oxides of S and Mo in the fumes from the kilns. The Mo can give rise to molybdenosis in livestock which is a pathological condition caused by excess Mo.

Fluorine is added to many public water supplies to improve dental health, especially to reduce the incidence of dental caries (which require fillings). In the US, 60% of potable waters have around the optimal concentration of F (1 μg/ml) either naturally, or by addition (Harte *et al.*, 1991). Tea leaves are a relatively rich source of F (and Al). In areas where local water supplies have a naturally high concentrations of F (> 1.5 mg F/l) people tend to have mottled teeth. Excessive intakes from water with > 3–6 mg F/l can cause skeletal fluorosis (deformed limbs) which becomes crippling with water concentrations > 10 mg F/l. However, in some places where this occurs, contamination of food (drying cobs of maize) by the smoke from cooking stoves burning high F-containing coal is the main source of F, rather than the drinking water (I. Thornton, pers. comm.).

5.5
Radionuclides

5.5.1   History and nomenclature

The first observation of radioactivity was made in 1896 by Becquerel, who noted that salts of uranium were phosphorescent and emitted radiation which penetrated black paper, opaque to light, and which reduced a photographic plate. For pure salts this activity is independent of the mode of chemical combination and of changes in temperature and magnetic flux. The activity always relates to the amount of the source element present and hence radioactivity is described as an infra-atomic property.

A principal ore of uranium is pitchblende which includes its black oxide U_3O_8. Marie and Joliot Curie with Bemont noticed that the intensity of emission varied with the source. Pitchblende from the Joachimstahl mine was four times stronger than that from Cornwall and both were more active than pure uranium nitrate. They concluded that these ores contained a further more active source of radioactivity; this led them to the discovery of polonium and then to radium, which was isolated as its sulphate and separated from the closely related barium salt by a patient fractionation of the chlorides. One ton of pitchblende afforded only 0.3 g of radium but its activity is over 10^6 times that of uranium.

5.5.2   Types of radioactive emission (Mellor, 1932)

The α-particles emitted by radium and uranium have the mass of helium atoms but carry two positive charges; one gram of radium emits 3.7×10^{10} α-particles and 2×10^9 cal, which is 2×10^5 times the calorific value of coal. The other principal source of radioactivity is the β-particle which is an electron carrying a single negative charge. Radioactive decay may give rise to other entities including the positron but these are not of concern as pollutants. The chemical consequences can best be shown by considering a decay series (Figure 5.12).

^{238}U is the predominant naturally occurring form, it loses 4 mass units and two nuclear charges on emission of an α-particle to give an isotope of thorium, written as ^{234}Th. This product decays with the ejection of a β-particle, which entails no mass change but a gain of one nuclear charge to form protactinium (Pa); a further β-emission leads to ^{234}U. This isotope contributes only 0.7% of natural uranium and this can be accounted for by

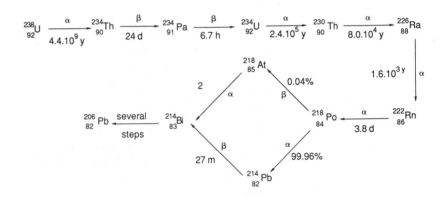

Figure 5.12   Decay of uranium 238 (Mellor, 1932).

considering the half-lives ($t_{1/2}$) of its predecessors. The two β-emitters have very short lives relative to that of ^{238}U, while they sustain the sequence through their activity, their rate of formation is less than their rate of decay and so only small amounts survive in the equilibrium. Because $^{234}_{92}$U decays much more slowly than $^{234}_{91}$Pa, it will accumulate to a relatively higher level than this precursor, nevertheless $^{234}_{92}$U with $t_{1/2} = 2.5 \times 10^5$ yr is shorter-lived than $^{234}_{92}$U by over 10^4 times and so survives at a lower concentration than the parent uranium.

The series is continued by the successive emission of two α-particles to produce ^{230}Th and then ^{226}Ra with eight fewer mass units than ^{234}Th and atomic number 88. Radium is in group II of the periodic table with chemistry similar to barium; loss of an α-particle gives $^{222}_{86}$Rn in the inert gas group zero. ^{218}Po in group VI is formed by further α-loss but this isotope has a dual decay path and yields the more electronegative ^{218}At by β-loss and also $^{214}_{82}$Pb by α-loss. Radioactive decay ends after several further steps with the stable isotope ^{206}Pb. There are two other natural decay series, one begins with ^{235}U and the other with ^{232}Th.

An important member of the ^{238}U series is the short-lived inert gas radon ($t = 3.8$ days) and owing to its release from granitic rocks we shall have to consider its toxicity within buildings erected on these foundations (p. 185).

γ-Radiation.   When a new species is produced by the emission of either an α- or β-particle it will be at an energy level above its stable ground state and to achieve stability the excess energy is emitted as γ-radiation.

5.5.3   Units of energy and measurement of toxicity

Energy is measured in electron-volts (eV). One eV is the energy acquired when unit electronic charge is accelerated through a potential difference of one volt and is equivalent to 1.60×10^{-12} ergs. The emission energies of radioactive particles are greatly in excess of this and are conveniently expressed in MeV or 10^6eV. Table 5.24 gives some representative values including those for a number of nuclides in the ^{238}U series. β-particles have a range which depends on the emission energy; for those of 0.5 MeV it is about 1 m and for those of 3.0 MeV it is about 10 m in air. When taken up by obstacles they produce a more penetrative secondary radiation known as bremsstrahlung. The best protective barriers are solids of low atomic number such as aluminium, perspex and rubber. 2.5 cm of perspex will protect against β-emission of energy as high as 4 MeV.

Exposure to radiation is measured in rads where one rad leads to the release of 100 ergs/g of body tissue. This is also equivalent to exposure to a level of one Roentgen.

Table 5.24   Representative emission
energies (Wilson, 1966)

Nuclide	Energy (MeV)	Half-life
β-*Emission*		
^{234}Th	0.19	24.1 days
^{234}Pa	2.31	1.18 min
^{210}Pb	0.063	22 years
^{35}S	0.167	87 days
^{36}Cl	0.71	3×10^5 years
^{40}K	1.32	1.3×10^9 years
α-*Emission*		
^{238}U	4.2	
^{234}U	4.77	
^{226}Ra	4.78	
^{222}Rn	5.48	

The current usage is to refer to exposure in rems. The rem takes account of a quality factor which for the more dangerous α-particles is 10, so that exposure to them at the level of one rad is rated as ten rems.

γ-rays are more difficult to stop because they are not ionic but their intensity falls away exponentially and a shield must be thick enough to reduce the intensity to acceptable levels.

Materials of high atomic number may exhibit photoelectric absorption whereby the energy of γ-rays is transferred to an electron, which is then ejected from the atomic shield. ^{60}Co is a dangerous product of nuclear fission (p. 171) and emits γ-rays of mean energy 1.26 MeV; to reduce their dose tenfold 4.6 cm of lead is needed and for each subsequent tenfold reduction a further 4.6 cm of shielding is necessary.

5.5.4   Radioactive potassium

Although not a member of a series potassium (^{40}K) presents a minor environmental risk (equation 5.40):

$$^{40}K \xrightarrow[t_{1/2}=1.3\times10^9 \text{ years}]{\beta} {}^{40}Ca \qquad (5.40)$$

In Europe and the USA ^{226}Ra occurs in concrete at an average level of 0.9–2.0 pCi/g (cf. p. 185), with ^{232}Th at 0.8–2.3 and ^{40}K significantly higher at 9–19 pCi/g. A 70 kg human contains 140 g of potassium, mostly in muscle, and the ^{40}K component emits 0.1 μCi and delivers 20 mrems to the gonads and 15 mrems to bone. Fortunately, although potassium salts are important and widely used, the proportion of the radioactive isotope is only 0.0118%.

5.5.5   Production of radionuclides by artificial means

Due in part to their industrial value as tracers many of these substances have been synthesized; they can be obtained by neutron bombardment. If there is a fruitful collision with a nucleus, loss of a proton or an α-particle will lead to chemical change while loss of γ-radiation will produce an isotope. An example of each is given below:

$$\text{n,p reaction } {}^{35}_{17}\text{Cl} + {}^{1}_{0}\text{n} \rightarrow {}^{35}_{16}\text{S} + {}^{1}_{1}\text{p} \tag{5.41}$$

$$\text{n,}\alpha \text{ reaction } {}^{27}_{13}\text{Al} + {}^{1}_{0}\text{n} \rightarrow {}^{24}_{11}\text{Na} + {}^{4}_{2}\text{He} \tag{5.42}$$

$$\text{n,}\gamma \text{ reaction } {}^{23}_{11}\text{Na} + {}^{1}_{0}\text{n} \rightarrow {}^{24}_{11}\text{Na} + \gamma\text{-rays} \tag{5.43}$$

${}^{35}_{16}\text{S}$ produced as in equation 5.41 is oxidised to SO_2 and is then injected into plumes to trace the extent of transport (p. 27). Its half-life of 87 days and strong β-emission make it well suited for this purpose and since it is chemically indistinguishable from unlabelled SO_2 its response to the environment is identical.

5.5.6   Nuclear fission

Discovery.   Irene and Frederic Joliot-Curie came close in 1934 when they noted that the product from neutron bombardment of ${}^{238}_{92}\text{U}$ was not the expected ${}_{89}\text{Ac}$ but was more like ${}_{57}\text{La}$. Hahn and Strassman, while attempting the synthesis of transuranium elements in 1938, were led to the first firm description of a fission reaction:

$$_{92}\text{U} + {}^{1}_{0}\text{n} \rightarrow {}_{56}\text{Ba} + {}_{36}\text{Kr} \tag{5.44}$$

In 1940 McMillan and Abelson observed that some deuterons recoiled with high energy resulting from fission when they were used to bombard uranium, but a new element was formed (equation 5.45) which lacked the energy to escape from the body of uranium. This was named neptunium (Np) after the first planet beyond Uranus:

$$_{92}^{238}\text{U} + {}^{2}_{1}\text{H} \rightarrow {}^{238}_{93}\text{Np} + 2\,{}^{1}_{0}\text{n} \tag{5.45}$$

Subsequently Seaborg showed that:

$$_{93}^{238}\text{Np} \xrightarrow[\text{2.1 days}]{\beta} {}^{238}_{94}\text{Pu} \tag{5.46}$$

This isotope of plutonium, named after the furthest planet, is an α-emitter with $t_{1/2} = 86$ years; that required for nuclear energy generation is ${}^{239}_{94}\text{Pu}$.

Criteria for fission.   Slow fission is possible but such events are rare and of no use for power generation, thus ${}^{238}_{92}\text{U}$ undergoes spontaneous fission to give ${}_{54}\text{Xe}$ which accumulates in natural sources and which can be used for dating.

E_{critical} is that required to split a nucleus in two. When this energy is less

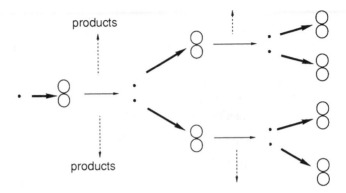

products

products

Figure 5.13   Ideal fission without neutron loss.

than the binding energy the atom only requires to capture a neutron of zero energy to undergo fission; it is said to be *fissile*. $^{235}_{92}$U is the only natural material with this property, capturing a neutron and undergoing fission as ^{236}U. The synthetic nuclide $^{239}_{94}$Pu is also fissile.

A nucleus is said to be *fissionable* if it reacts only after capturing an energetic neutron. In order to sustain an energy-generating chain reaction in fissionable nuclei one energetic neutron must survive to initiate the next generation. In practice more than one such neutron must be produced on fission since some escape from the system and others are captured by accumulating non-fissionable products. Figure 5.13 shows the rapid process which follows the release of two neutrons leading to four reactions in the second generation.

The availability of energetic neutrons is expressed as the *multiplication factor* (k) = number of fissions in one generation/number in the preceding generation. Provided that $k > 1$ the conditions for maintaining a chain reaction will persist and be enhanced with each generation. This *super-critical* state was necessary in the $^{235}_{92}$U bomb which was detonated over Hiroshima on 7 August 1945.

When $k = 1$ the system is said to be *critical* and this is the state required for the controlled release of energy.

When $k < 1$ the system is *sub-critical* and unproductive.

Quantitative yield.   To appreciate the potential of fission for energy production it is helpful to analyse a typical reaction:

$$^{235}_{92}U + ^{1}_{0}n \longrightarrow ^{95}_{42}Mo + ^{139}_{57}La + 2\,^{1}_{0}n + \beta\text{-particles} \quad (5.47)$$

235.0439 1.0087   94.9057   138.9061 2.0174

236.0526   235.8292

By subtraction a mass loss Δm of 0.2234 units is obtained. To obtain the energy equivalent of one atomic mass unit the reciprocal of Avogadro's number is put into the Einstein formula $E = mc^2$, where c is the velocity of light, 3×10^{10} cm/s.

Hence mu energy $= 1/(0.6 \times 10^{24}) \times 9 \times 10^{20} = 1.5 \times 10^{-3}$ ergs

Using the conversion factors this can be expressed as:

$$1.5 \times 10^{-3} \times 10^{-7} = 1.5 \times 10^{-10} \text{ Joules or as}$$

$$1.5 \times 10^{-3} \times 6.24 \times 10^{5} = 931 \text{ MeV}$$

It follows that 1 g of $^{235}_{92}$U will produce

$$931 \times 0.6 \times 10^{24}/235 = 2.38 \times 10^{24} \text{ MeV}$$

5.5.7 Power generation in nuclear reactors

In this book only a brief summary is given; for a more detailed treatment see Nero (1979) and Lamarsh (1983).

Preparation of the fuel.

Gaseous diffusion The formidable problem of enriching the low level (0.7%) of $^{235}_{92}$U in natural uranium was initially dependent on conversion of the oxides to uranium hexafluoride. Then the corrosive mixed gases were passed through nickel columns with pores of 0.01 μm when the lighter ^{235}UF$_6$ escaped faster than ^{238}UF$_6$; since the fractional separation was only 1.002 a multi-stage process was needed. After enrichment the hexafluoride was converted to the dioxide by hydrolysis:

$$\text{UF}_6 + \text{steam} \rightarrow \text{UO}_2.\text{F}_2 \xrightarrow{\text{H}_2} \text{UO}_2 + 2\text{HF} \qquad (5.48)$$

Alternatively the tetrafluoride may be reduced to uranium metal (Greenwood, 1984):

$$\text{UO}_2 + 4\text{HF} \underset{-\text{H}_2\text{O}}{\overset{550°C}{\rightarrow}} \text{UF}_4 \underset{700°C}{\overset{\text{Mg}}{\rightarrow}} \text{U} + 2\text{MgF}_2 \qquad (5.49)$$

Centrifugation A more economical process is to place the UF$_6$ in a series of gas centrifuges where now the heavier isotope migrates more readily to the boundary and ^{235}UF$_6$ becomes concentrated near the axis of rotation.

Laser separation This depends on the raising of uranium metal to an excited state rather in the manner of AA spectroscopy (p. 88). A laser emitting green light can be tuned precisely to a wavelength of 502.73 nm so as to excite only ^{235}U. It is then ionized by a second source and electromagnetically separated from residual uncharged ^{238}U. In practice much of the original uranium vapour is also collected and so the net change is one of enrichment and not complete separation.

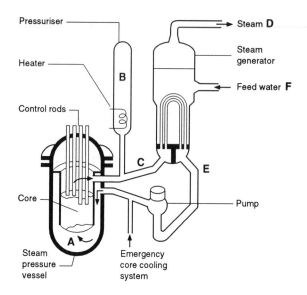

Figure 5.14   Simplified diagram of a PWR (Collier, 1987).

5.5.8   Nuclear reactor types

The pressurized water reactor (PWR).   This is widely used today around the world. Ordinary water is used as a moderator for the production of low energy or *thermal* neutrons, that is those with energy comparable to the system as a whole. Water is favoured because it is cheap and being composed of lighter atoms it is efficient. In Figure 5.14 the oxide fuel containing 3% $^{235}UO_2$ is supplied as 1 cm pellets packed within 3.5 m long hollow zircalloy tubes or pins, which are grouped into bundles or elements of about 20 cm^2. The reactor (A) contains about 200 of these elements with spaces for control rods, coolant flow and neutron detecting instruments. Moderating water enters at 290°C and leaves at 320°C and hence the reactor chamber is at a high pressure of about 2000 psi or 15 MPa (see appendix). The containing steel vessel is some 12 m high and 5 m in diameter.

The pressurizer (B) is fitted to ensure that there is no loss of coolant by evaporation; this would be especially serious as neutron energies would also rise wherever steam replaced water.

The issuing hot water (C) is passed through a heat exchanger to produce steam which leaves (D) to drive the turbine. After exchanging its heat the cooling water is returned at (E), while fresh cooling water and turbine condensate enter the exchanger at (F). The primary coolant in contact with the reactor is isolated from the secondary heat exchanger circuit.

The multiplication factor (p. 172) may be rapidly varied by movement of the control rods which are withdrawn to raise the power level and inserted to reduce it. If need be they can shut the reactor down completely. The rods are neutron absorbing materials such as boron steel or silver alloys and their action is supplemented by injection of boric acid solution into the moderator.

The boiling water reactor (BWR).   At pressures of about 7 MPa the reactor can be stabilized even when the moderator boils and the emitted steam can be used directly to drive the turbines. However, these and all the linking condensers and pipework must also be shielded as the circulating water becomes radioactive.

The advanced gas cooled reactor (AGR).   This was developed originally in England and the reactor at Hinkley Point B is of this type with enriched UO_2 fuel in steel jackets. The moderator is constructed from graphite bricks and CO_2 is circulated to cool the system and to generate steam. Although this reactor operates at an efficiency of 40% it does not produce electricity more cheaply than a PWR which has been the preferred design in the UK since 1979.

The heavy water reactor.   The Canadian design is the CANDU, an acronym for CANada and Deuterium Uranium. It was developed there after World War II to take advantage of available natural uranium without the need for enrichment; this is because heavy water has a low absorption for thermal neutrons. It also follows that to reach thermal energies energetic neutrons must undergo more collisions and so travel further in D_2O than in light water.

A simplified diagram (Figure 5.15) shows the fuel rods passing through a tank, or calandria, containing the moderator at a low pressure while a high coolant pressure is maintained only within the fuel channels. This has the advantage of limiting the volume of the vessel but it also means that the steam pressure and temperature are less than in the PWR and the efficiency is only 30%.

Breeder reactors – the liquid metal fast breeder reactor.   This one type (LMFBR) will be considered to illustrate the principle. These reactors are attractive to nations with no indigenous fuel supply because they produce fissile material in the course of their operation and can use depleted fuel from thermal reactors.

Figure 5.16 broadly outlines a typical arrangement. Here the driving force is obtained from fast neutrons and there is therefore no moderator; the cooling and heat transfer circuits are charged with metallic sodium (m.p. 98°C). From a purely chemical point of view this is a hazardous

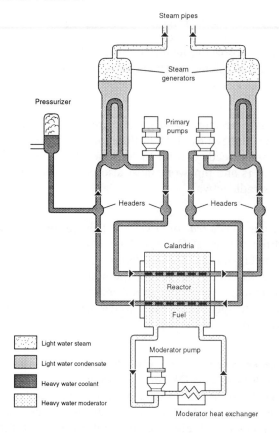

Figure 5.15   Diagram of a CANDU reactor (Nero, 1979).

choice because of its high reactivity to water and oxygen and the container must be flooded with nitrogen gas to limit these risks. However, it has engineering advantages in that sodium has only a slight moderating effect and at the operating temperature of 500°C a high pressure vessel is not required. In the breeder reactor sodium within the inner cooling circuit captures some neutrons to form the higher nuclide:

$$^{23}_{11}\text{Na} + ^{1}_{0}\text{n} \rightarrow ^{24}_{11}\text{Na} \tag{5.50}$$

Hence a double looped system is required to prevent circulation of active material outside the screen.

If a breeder reaction is to be sustained, the ratio of fissile atoms produced to the average number of fuel atoms consumed must be > 2. This is because of the demands: one neutron is required to sustain the critical state, one neutron is required to produce fissile material and some excess is necessary to make up losses.

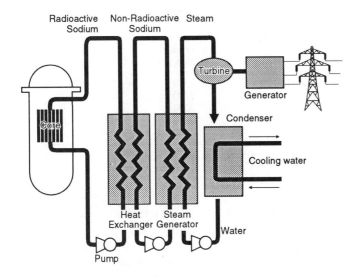

Figure 5.16 Diagram of a fast breeder reactor (Nero, 1979).

Under typical conditions the reactor will be charged with ^{238}U mixed with about 20% by weight of ^{239}Pu. When ^{239}Pu captures a neutron of essentially zero energy the product ^{240}Pu has binding energy of 6.4 MeV which is in excess of the energy required for fission (4.0 MeV) and so the self-sustaining reaction begins. In a year of operation about 50 times the initial weight of ^{239}Pu will be produced (Lamarsh, 1983) together with other products (Table 5.25).

Table 5.25 Some fission products

Element	Symbol	$t_{1/2}$	Activity $\times 10^3$ Ci/ton uranium	Risk level
Strontium	^{89}Sr	50 days	93	2
	^{90}Sr	28 years	92	2
Ruthenium	^{106}Ru	1 year	760	2
Caesium	^{134}Cs	2.1 years	203	2
	^{137}Cs	30 years	130	2
Iodine	^{129}I	1.7×10^7 years	low	
	^{131}I	8 days	72	
Plutonium	^{238}Pu	86 years	5	
	^{239}Pu	2.4×10^4 years	low	
	^{240}Pu	6.5×10^3 years	low	1
	^{241}Pu	13 years	175	
	^{242}Pu	3.8×10^5 years		

Ci = Curies

Table 5.26   Spread of ^{137}Cs from Chernobyl (Nuclear Energy Agency, 1989)

Country affected	Distance from Chernobyl (Median km)	Activity* ^{137}Cs**	^{131}I
Austria	1250	23	120
Norway	2000	11	77
UK	2250	1.4	5.0
France	2000	1.9	7.0
Ireland	2750	5.0	7.0

* k Bequerels/m^2 (p. 185)
** with ^{134}Cs

Strontium 90 presents a risk inherent in its relatively long half-life and its chemical affinity with calcium, which leads to its concentration and persistence in bone; particulate emission also presents a hazard to lungs. After the Urals disaster in 1958 ^{90}Sr was detected in soil at levels up to 43 curies/ha and it was distributed up the food chain in plants and small mammals (Medvedev, 1979).

Caesium 137 is similar to potassium in its chemistry and therefore it penetrates neural tissue and presents a hazard to muscle (Adelman, 1987).

Release by leakage from the Windscale (now Sellafield) plant of British Nuclear Fuels peaked at 1.2×10^4 C/month in 1975 (Crouch, 1986). Fish caught in this part of the Irish Sea were marketed with ^{137}Cs levels of 10 pCi/g compared to a typical level in North Sea catches of 0.1 pCi/g.

In the aftermath of the Chernobyl incident concern arose from the spread of ^{137}Cs and the levels at some distant sites are given in Table 5.26.

Iodine 131 is combined and concentrated in the thyroid gland following the ingestion of milk from cows grazing on polluted pasture, as was the case following the Windscale release of 1957.

Ruthenium 106 has been found in seaweed at levels of 200 pCi/g – a concentration factor of 2000 with consequences for those who consume laverbread.

Plutonium nuclides do not follow a natural path in a food chain as they are artificial, nevertheless ingestion of their water-soluble forms presents a hazard to bone and to liver. Insoluble forms inhaled as dust induce lung cancer.

5.5.9   The future of nuclear power

The first successful generator of power for the public supply came on stream in 1956 at Calder Hall, Cumberland. This stimulated international enthusiasm and massive capital investment. One motivating factor in the climate of the 'cold war' was the production of plutonium for nuclear weapons. Today a more sober attitude is taken largely as a result of the public perception of risk of exposure to radiation and also the properly critical assessment of cost of nuclear generated electricity. Although other resources of fossil fuels have been identified, shortages of power and serious social consequences will be experienced in the long term if no other alternative is available. While energy-demanding research into fusion energy is continued there remains a pressing need for energy conservation. In its report of 1981, the House of Commons Select Committee on Energy observed that 'The Department of the Environment has no clear idea as to whether spending $1300 million on a new nuclear plant is as cost effective as spending the money on energy conservation'.

Risk of exposure to radiation.   The International Atomic Energy Agency based in Vienna has proposed a system of classification for nuclear accidents (Table 5.27) on a scale from 4 to 7, with less severe incidents on scale from 1 to 3.

Owing to the perceived risk nuclear stations are built as far as possible from major conurbations at sites where large quantities of cooling water are available. The distribution of stations in the USA (Figure 5.17) and the UK (Figure 5.18, at 5 times the US scale) shows that distance affords more protection to the former community than it does to the latter.

The nuclear industry recognizes the concept of 'maximum acceptable risk' based on an assessment of probabilities used originally in the chemical industry (Jackson, 1992). This, for example, put the risk of death from exposure to radiation in the UK at 1 in 10^7, the same as that from a lightning strike. These evaluations are necessarily based on notional data and not on experimental facts. They do not take account of the uniquely widespread effects and long term pollution arising from nuclear accidents. The uneasy public perception will inevitably be based on the evident consequences of a few major accidents rather than the purely statistical approach.

Table 5.27   Some nuclear events as placed on the IAEA scale of risk

Level	
7	Chernobyl, USSR, 1986
6	Ural mountains waste explosion, 1958
5	Fire at Windscale (Sellafield), 1957
	Three Mile Island, USA, 1979
4	Fatal accidents at Los Alamos, Wood River and Idaho Falls in period 1945–1964 and Saint Laurent, France, 1980
3	Unauthorised release at Vandellos, Spain, 1989; Tomsk-7, Russia, 1993
2	Incidents necessitating safety reassessment at UK Magnox stations, 1968, 1983, 1989
1	Management deficiencies in waste reprocessing, Windscale, 1986

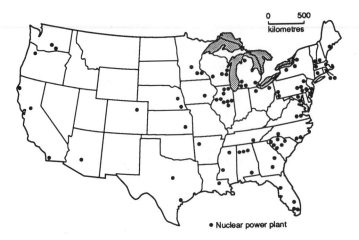

Figure 5.17   Distribution of nuclear stations in the USA (*Nuclear News*, 1992).

5.5.10   Observations on major accidents

The fire at Windscale (8 October 1957).   This occurred in the number one plutonium production reactor and was the result of human error coupled with inadequate operating instructions (Patterson, 1976). The reactor required purging by first raising and then lowering the power, but it appeared when the temperature was falling that the operation was incomplete. The crucial error was that power was boosted again to raise the temperature, however that indicated was not maximal and hotter spots elsewhere in the core took fire. No sign of malfunction was evident until 10 October when radioactivity was detected at the top of the massive discharge stack.

In the interim a major fire had taken hold and was consuming the uranium metal fuel and the graphite moderator. The operators attempted to maintain secrecy for 24 hours after the fire was discovered but as the supply of carbon dioxide available to them only fuelled the fire they had to involve off-site fire brigades who quenched it with water injected through fuel entry ports.

It was estimated that the most hazardous release was that of 20 000 Curies of ^{131}I (see Table 5.26) and the Government ordered the disposal of all milk from dairy herds within a radius of about 25 miles. The full report of the subsequent enquiry was never published.

The Urals disaster (1958).   This affected an area of some 1000 square miles between Cheliabinsk and Sverdlovsk causing death and destruction of property. An explosion occurred in waste from a plant which was most

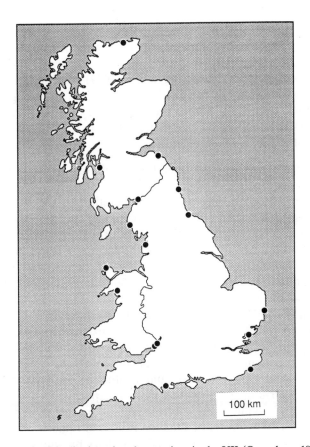

Figure 5.18   Distribution of nuclear stations in the UK (Openshaw, 1986).

probably producing weapons grade plutonium. Many years elapsed before any information reached the West but a full account by Medvedev was published in 1979; his early advice was rejected as 'rubbish' by the then Chairman of the UKAEA, Sir John Hill.

Three Mile Island (28 March 1979).   The reactor was operated by Metropolitan Edison on the Sasquehanna river close to the small community of Middletown and 7 miles from Harrisburg (population 60 000) and about 75 miles from Baltimore and Philadelphia (see Figure 5.19).

It is worth noting that the Atomic Energy Commission experts who published the accident probability assessment WASH-740 (*Theoretical Possibilities and Consequences of Major Accidents in Large Nuclear Power Plants*, 1957) were themselves of the opinion that a medium-sized accident would occur in their own lifetimes. However, no measures were in place to protect the local community, in the event the company informed the

Middletown authorities that 'Even the worst possible accident postulated by the AEC would not require evacuation of the town'.

The official report into the accident (*The Report of the President's Commission on 3 Mile Island*, Pergamon, 1979) concluded that operator error was responsible but that inadequate training, confusing procedures and failure to appreciate earlier accidents were factors. A review of the incident has been published (Stephens, 1980) and the following abridged sequence of events is taken from it:

1. 31 December 1978: plant operations begin just in time to earn a $40 million tax credit for the year.
2. 28 March 1979: during a routine operation to renew clogged ion-exchange resin in one of the large columns used to remove minerals from the feedwater, a small volume of water was forced into the compressed air control system for the facility. After a time-lapse this led to failure of one of the feedwater pumps and stopped the circulation of cooling water. Up to 1000 gallons/min could be supplied to the whole system.
3. Within seconds the temperature and pressure in the reactor rose rapidly. A release valve ejected a blast of excess steam to the open air.
4. The boron and silver control rods functioned and dropped amongst the 37 000 zirconium clad UO_2 fuel pellets so closing down the reactor. Pressure fell to a safe level but continuing radioactive decay still generated 6% of the operating thermal capacity.
5. Owing to the fault in the control system a further valve failure led to the release of a quarter of a million gallons of radioactive coolant onto the floors of the containment and subsidiary buildings.
6. Still only 1 minute into the incident the 1 m bore coolant inlet pipes hammered as pumps tried to force water past input valves closed for a previous test and *not reopened*. Indication of the closure was given on the control panel but was overlooked. Owing to the fault in the control system very many visual and oral alarms were activated at the same time so confusing the engineers.
7. The engineers were unaware of the loss of coolant and concluded that the problem was the result of flooding of the pressurizer (this keeps the pressure close to the optimum either by spraying water to condense excess steam or by evaporation of excess water by electrical heating). Hence more water was drained and the second flow pump was cut. The probability of failure of both feedwater valves had been put at 1 million to 1 against.
8. Steam bubbles forced out the remaining water. The overall temperature was recorded as 400°C but some thermocouples registered 1100°C and were disbelieved – here meltdown was occurring.
9. The intense radiation converted steam into hydrogen, which accumulated dangerously, and oxygen which reacted with the zirconium

cladding so loosing the UO_2 pellets inside the reactor. Under the conditions the silver and boron control rods were all destroyed. Fortuitously the release of the iodine nuclides was limited by combination with this silver to give involatile silver iodide.

10. 2 hours into the incident radiation at a level of 1000 millirems was detected in an auxiliary building.

11. On the Friday following the incident a helicopter at 600 feet above the station recorded activity at 1200 millirems/hour – 4 hours at this level would be equivalent to the EPA approved dose for a worker for a year. Owing to a failure of communication this activity was held to be that at ground level off-site. In the ensuing panic it is estimated that 40 000 people fled the area although no official order was given to evacuate.

In conclusion it should be noted that although the core came close to meltdown no one was physically injured; the major health effect was one of mental stress. There was, however, the loss of the reactor and the cost of clean-up, put at \$2 to \$3 billion levied on the taxpayer. 2000 truckloads of waste were transported across Washington State and 4000 gallons of polluted water were diluted and discharged into the river. Bruce Babbit, in a minority view to the report of the President's Commission, queried the running of nuclear plant by the private sector.

Chernobyl (26 April 1986) (OECD, 1987; Marples, 1986). The town dates from the twelfth century and lies in a sparsely populated area 135 km north of Kiev. The generating complex was active in October 1977 and by 1986 four 1000 MW units were producing 10% of the total USSR capacity.

The accident was the result of operator error while conducting an experiment in obtaining back-up power from the spinning turbines after shut-down. This was necessary because back-up diesel power for the pumps took 50 s to maximize. The release of information by the authorities was said to have been made with the principal aim of preserving a nuclear capacity in the country.

During the experiment the operator failed to set the power level control, this led to steam condensation and loss of productive neutrons. While purging the system of xenon, which acts as a moderator, more loss of neutrons occurred and *contrary to the safety procedure* the control rods were withdrawn and the safety system was closed off in order to complete the experiment against the clock.

During 4 s the power output rose from 200 MW to over 300 000 MW. Similar disruption of the fuel took place as at Three Mile Island and under the pressure of hydrogen, steam and burning graphite the 1000 ton top plate was blown off. Radioactive steam was distributed throughout the whole plant via the ventilation system and 300 on-site workers were

Figure 5.19   Three Mile Island nuclear power station (the damaged reactor is in the building on the left adjacent to the cooling towers) (photo: B.J. Alloway).

treated for radiation poisoning, while measuring devices for firemen went off-scale.

The public demand for information could not be resisted in the light of reported radioactivity 800 miles away over Sweden. Nevertheless the responses tended to vary with the national interest – France with 65% of nuclear generating capacity said nothing for 14 days. In the UK, Secretary of State Kenneth Baker told the House of Commons 'If there is no further discharge the effects will be over within a week', but two months later on June 20th grassland in the Lakes, North Wales and Scotland was so polluted that one million sheep and lambs were too radioactive to be slaughtered for food (see Table 5.26).

Six years after the accident children in the region of Kiev are being born with birth defects including adrenal and thyroid cancers. The Ukrainian authorities put the number of radiation related deaths at several thousand with 15 000 affected by disease to a lesser extent.

5.5.11   Radioactive release within buildings

Notwithstanding the hazards posed by occasional failure of control of nuclear power stations, the penetration of buildings by chemically inert radon gas presents a far more serious risk to human health (National

Academy, 1981). As is so often found, the risk to humans was first in evidence amongst industrial workers. Increased incidence of lung cancer in miners in western USA and Czechoslovakia was attributed to their inhaling ^{222}Rn.

^{222}Rn is a product of the decay of ^{238}U (Figure 5.12) and emanates from radioactive minerals in building foundations. Because it has a half-life of only 3.8 days the risk is only significant when the gas can quickly rise into the immediate subsoil, conditions which obtain above fissured granite. In the UK high levels occur in Cornwall, Devon, Derbyshire, Northampton-shire and in Scotland; New Yorkers living in basements and lower floors may also be affected. The pressure within buildings is normally lower than without and so radon enters via cavities and cracks in walls and through construction joints; it has seven times the density of air and so basements and ground floor rooms are the most exposed. Other nuclides of radon exist but they do not present the same risk as that from ^{222}Rn, for example ^{220}Rn formed by decay of ^{232}Th has too short a half-life to allow survival above ground.

The worst effects arise from *solid* α-emitting daughter elements which associate with moisture and dust and become deposited in the lungs. Reference to Figure 5.12 shows that these include ^{218}Po ($t_{1/2}$, 3.1 min), ^{214}Pb ($t_{1/2}$, 27 min) and ^{210}Bi ($t_{1/2}$, 5 days). Even those nuclides of short half-life will have an effect if they are being renewed continuously from leakage of radon into the air space.

Units. The ^{222}Rn dose is measured in Becquerels (Bq, p. 167) which is defined as 1 disintegration/s, a tiny fraction of the Curie (Ci, p. 170) which is equivalent to 3.7×10^{10} dps. For practical purposes the pico-Curie is preferred.

The median level outdoors in New York is:

$$180 \text{ pCi/m}^3 = 180 \times 10^{-12} \times 3.7 \times 10^{10} \text{ Bq/m}^3 = 6.7 \text{ Bq/m}^3$$

The median in the basements and ground floors of buildings in New York is 30 Bq/m^3, evidence of greater risk indoors (cf. p. 220).

In 2100 homes in the UK the average concentration is 20 Bq/m^3, with most below 100 Bq/m^3 but including 48 above this level (Gardner *et al.*, 1992). The National Radiological Protection Board advised that remedial action be taken when levels exceed 400 Bq/m^3. This degree of exposure corresponds to a 2% risk of death from lung cancer in non-smokers, a relatively small proportion of the risk of lung cancer run by smokers and put at 20% overall.

Remedial measures. These include sealing cracks in structures and laying building weight polythene over floors. A more expensive measure is the construction underfloor of a low pressure sump packed with porous material and vented outdoors by an extractor fan.

Table 5.28 Proportion (%) of power generated by
nuclear fission in some developed nations

Belgium	60	West Germany	31
Sweden	43	Spain	22
UK	21	France	65
USA	15		

5.5.12 Social aspects of nuclear power generation

The principal factors which weigh against an expansion of nuclear power
are:

1. The recorded major accidents.
2. Doubts about the safe ultimate disposal and reprocessing of radioactive
   waste as evidenced by the THORP facility.
3. Lack of confidence in Government agencies as a result of concealment
   and denial of information.
4. Inaccurate accounting and the recent admission that nuclear electricity
   is not cost effective.

Nevertheless the developed nations depend to a significant extent on
nuclear stations for their electricity supplies (Table 5.28). Italy and
Switzerland have halted nuclear development. Should other nations with
intense urban and industrial development follow suit, considerable construction of new fossil fuel units would be required. Their building costs
would rise because of the need to control CO_2 emissions. Meanwhile
China, France and Spain continue to build; Japan is aiming for 40 nuclear
units by 2010 and to be 40% nuclear by 2050. Czechoslovakia has eight
stations and plans a further eight although there are doubts about the
safety of their plant at Bohunice, which is close to the Austrian border.
Similarly North Korea plans 4 reactors at Sinpo, 300 miles from Vladivos-
tok, and has rejected supervision by the Atomic Energy Authority. This
still leaves the risk of pollution from other suspect sources, for example
there are 15 PWR stations of the type existing at Chernobyl that remain in
use in Eastern Europe.

It is not possible to enlarge on the engrossing history of nuclear power in
this book but the reader is referred for details to Hall (1986), Pocock
(1977) and Patterson (1985).

5.5.13 Power from thermal fusion

The great density and high temperature of the sun's plasma leads to a
continuous release of energy through the combination of protons. This
occurs in a controlled way rather than as a massive explosion as the

probability of the reaction is low. This depends on one of the colliding protons shedding its positive charge as a positive electron or positron, energy release then occurs through reaction of the residual neutron with its partner proton:

$$\gidefault{}_0^1n + {}_1^1H \rightarrow \qquad {}_1^2H \qquad + \text{energy} \qquad (5.51)$$
$$\text{a deuteron (d)}$$

The new deuteron will react rapidly with one of the many protons available in the dense material:

$${}_1^2H + {}_1^1H \rightarrow {}_2^3He + \gamma\text{-rays} \qquad (5.52)$$

In the sun energy release occurs through the sequence:

$${}_2^3He + {}_2^3He \rightarrow {}_2^4He + 2\,{}_1^1H \qquad (5.53)$$

The two protons formed re-enter the cycle, with ${}_2^4He$ providing more energy by combinations such as:

$${}_2^4He + {}_2^3He \rightarrow {}_3^6Li + {}_1^1H + \text{energy} \qquad (5.54)$$

Terrestrial experiments on thermal fusion are of a similar type but differ in detail. Two atoms of deuterium may combine with similar probabilities to produce helium or tritium:

$${}_1^2H + {}_1^2H \rightarrow {}_2^3He + {}_0^1n + 2.45\ \text{MeV} \qquad (5.55)$$

$${}_1^2H + {}_1^2H \rightarrow {}_1^3H + {}_1^1H + 3.0\ \text{MeV} \qquad (5.56)$$

These energy outputs are about one million times that obtained from a typical chemical change and require an input of only a few thousandths of that produced.

The reaction between deuterium and tritium is even more productive:

$${}_1^2H + {}_1^3H \rightarrow {}_2^4He + {}_0^1n + 18\ \text{MeV} \qquad (5.57)$$

but tritium has disadvantages as a source in that its production is expensive in terms of capital and of the energy input required.

Apart from the low level of radioactivity of tritium the fusion processes avoid the hazards inherent in nuclear fission. This gives it a degree of priority which is heightened by public fears and political problems over the extension of existing nuclear fission programmes. Thermal fusion faces formidable experimental difficulties if a continuous gain is to be won.

Although only a few keV are needed to bring two nuclei together, for continuous production the fuel must be of a density high enough to generate sufficient collisions and exist at temperatures in excess of that in the sun's plasma. The destructive body of gas is contained *in vacuo* out of contact with the reaction vessel within a torus of magnets. The critical level requires that a temperature of the order of 10^8 degrees and density of 10^{20} ions/m^3 are maintained for one second. These conditions have been met singly in the Joint European Torus (JET) but not in a concerted manner.

5.5.14   Cold fusion

The formidable difficulties encountered with hot fusion encouraged the search for an alternative based on an idea of Frank in 1947, which depends on the effect of muons on atomic properties. A muon bears a single electronic charge but has 207 times the mass of the electron. On interaction with deuterium or tritium the muon enters an orbit much closer than that of the electron, which is ejected since it is no longer attracted by the balanced charges. The resulting atoms of deuterium or tritium are reduced in size by the factor of 207 and resemble neutrons. This change greatly enhances thermal fusion but unfortunately has yet to produce more energy than that required to generate a muon in a particle accelerator. The lifetime of a muonic atom is only a few microseconds which limits the number of collisions that it can catalyse.

Close (1990) gives a fuller account of fusion processes and of the history of events, including the recent mistaken claim that fusion occurred when electrolytically generated deuterium becomes concentrated in a palladium electrode.

5.6
Mineral fibres and particles

5.6.1   General aspects

Mineral pollutants, as classified here, enter the environment as fibres or particles of crystalline inorganic compounds (frequently as silicates) of relatively fixed chemical composition. In most cases they will remain as fibres or particles and exert their potentially harmful effect in this form, although all will undergo chemical weathering at rates determined by the mineral, and the environment. For example, quartz particles will be very resistant to weathering but calcite particles will rapidly dissolve in moist acid environments. Other mineral pollutants, such as metallic ore minerals, whose major toxic effect is caused by the release of metal ions on weathering in either water or soil, are dealt with elsewhere in the book (section 5.3).

Mineral fibres are commonly used for the thermal insulation of buildings, pipes and storage tanks, fireproofing, reinforcement of construction materials. Examples of uses include glass fibre reinforced plastics (GRP or 'fibreglass') and asbestos reinforced cement products including roofing panels and pipes for drinking water. Asbestos withstands the high temperatures caused by friction and is used in the manufacture of brake linings for motor vehicles and is also woven into cloth for use in fire protection suits. Rock wool is used as cavity wall thermal insulation (blown into the cavity) and as a rooting medium for hydroponically grown plants.

Unlike most other pollutants reviewed in this book, the health hazard from mineral fibres and particles is not due to a biochemical toxic reaction but to an irritational effect related to the size, shape and surface of the particles which causes an inflammatory reaction in body tissues (especially the lung) and the formation of scar tissue or even a carcinoma. Inhalation of separate dry fibres in a confined air space is the major hazard. When the fibres are bonded they constitute a relatively low hazard because there will be few loose fibres. However, damaged thermal insulation (e.g. during alterations or demolition), and accumulated fibres from the manufacture or wear of asbestos or other mineral fibre materials (e.g. brake shoe/pad dust) constitute a major hazard. Appropriate safety masks and skin protection should always be worn when handling mineral fibres.

Dust particles can range in size up to 150 μm but those with diameters (or lengths) of <10 μm are more hazardous to health because they tend to remain suspended in air for long periods and those smaller than 5 μm can penetrate the respiratory system reaching the bronchioles and alveoli of the lungs. Mineral fibres have an aerodynamic configuration which results in them being very easily suspended and resuspended in air and therefore exacerbates their hazard to health.

The trachea and larger bronchi are lined with cilia (fine hairs) which help to filter inhaled air and transport particles removed from this air in mucus carried up and out of the respiratory tract by coordinated beating of the cilia. However, although no longer a potential hazard to the lungs, these particles can still affect health when they are swallowed into the stomach in mucus.

Many inhaled dusts accumulate in the lungs without stimulating any local reaction and therefore do not cause any recognizable disease. Urban dwellers commonly have accumulations of dark material in their lungs. Benign (non-cancerous) pneumoconioses are recognized as being caused by prolonged exposure to $CaCO_3$, $CaSO_4$, Fe, Sn and Sb dusts.

5.6.2 Analysis

Unlike most pollutants, mineral particles are easily seen with a microscope and counted on air filters. Identification is usually by X-ray diffraction spectrometry (XRD). Mineral pollutants are difficult to detect in soils and sediments owing to the abundance of similar minerals although scanning electron microscopy with linked X-ray analysis can be used to identify the particles.

5.6.3 Examples of mineral pollutants

Asbestos. This is the group name applied to a range of naturally occurring fibrous magnesium silicate minerals with an approximate formula of $Mg_3P(Si_2O_5)(OH)_4$. Three common types are used:

chrysotile (white asbestos) – least hazardous to health;

crocidolite (blue asbestos) – most hazardous;
amosite (brown asbestos) – second most hazardous.

Sources in the environment.   These mainly comprise mining for asbestos
and other magnesium silicate minerals, docks handling the raw material,
factories making asbestos products, demolition of buildings (pipe and
boiler insulation, roofing materials), ship breaking, vehicular brake linings
(especially in confined spaces such as underground stations and motor
repair garages), waste disposal and, much more insidiously, particles
carried on clothing (which can result in the exposure of asbestos worker's
families).

Toxicity.   Asbestos fibres can cause lung and bowel cancer as well as
non-cancerous lung diseases. Approximately 50% of the inhaled fibres are
cleared from the lungs and swallowed which then exposes the throat and
digestive system to their hazardous effects. Water from asbestos-cement
pipes also poses a further source of digestive tract exposure. Respiratory
diseases include asbestosis (a pneumonia-like condition) bronchial cancer
and mesothelioma (with a latency period of 20–30 years). There appears to
be a synergistic reaction between cigarette smoke and asbestos in that the
onset of disease is more pronounced in heavy smokers than in non-
smokers.

The widespread use of asbestos in schools and other public buildings for
thermal insulation and decorative materials has resulted in very large
numbers of people being potentially exposed to these particles. The
transport and dumping of building wastes has also contributed to the
amounts of asbestos particles in the air. In the USA it has been estimated
that the major emissions of asbestos into the air are from mining and
milling of asbestos (700 t/yr), product manufacture (100 t/yr) and landfills
(18 t/yr). The current US limits in workplace air are 2 fibres/cm^3, but
stricter limits of 0.5 or 0.2 fibres/cm^3 are proposed (Harte *et al.*, 1991).

Non-asbestos mineral fibres (such as glassfibre and rockwool) are
increasingly used as substitutes for asbestos. While they are not as
hazardous as asbestos, excessive exposure to inhaled fibres could still cause
disease.

Silica mineral particles

Source.   Quarrying, rock crushing, ceramics.

Toxicity.   Fine particles of quartz and other forms of silica (SiO_2) can
accumulate in the lungs and cause the formation of localized nodules of
scar tissue (fibrosis) which enlarge, merge and reduce the respiratory
function of the lungs. This disease is called silicosis and it has a long latency

period so it could be well advanced before the symptoms are recognized. Silicosis is frequently associated with infections such as bronchitis and even tuberculosis.

Coal dust.   Coal is a complex organic polymer containing bound trace metals (<12% inorganics) sometimes with sulphide mineral inclusions. It has a high calorific value (*c.* 28 MJ/kg) and therefore is a combustion or explosion hazard (see section 6.2).

Source.   Mining, transport and handling of coal.

Toxicology.   Prolonged exposure to coal dust produces a fibrosis condition of the lung called pneumoconiosis which is distinguishable from silicosis because the lung lesions are black. Lung tissue is obliterated and becomes prone to recurrent infection causing severe respiratory disability. This disease is a major occupational health hazard of coal mining.

Other particulate pollutants.   Respirable fine particles (<5 μm) of several metals or their compounds including Be, Ni, and Cr can constitute serious hazards to health through their effects on the respiratory system. However, these mainly affect occupationally exposed workers or possibly people living near to industrial plants using or producing the metals or their compounds. These metallic pollutants are dealt with in section 5.3.

Many ubiquitous particles which are commonly found suspended in air can cause respiratory illness in susceptible people. These particles and fibres include pollen, mould spores, faeces and pieces of exoskeleton of the house dust mite (*Dermatophagoides pteronyssinus*), animal hairs and skin fragments (dander) and bird feathers, all of which can cause allergic conditions, such as asthma and hay fever (Rowland and Cooper, 1983). There is considerable interest in this subject nowadays owing to the occurrence of the 'sick office' syndrome in buildings with large air-conditioned offices. Likewise, asthma incidence is increasing in the general population and may be due, at least in part, to the increasing amount of carpeting used in houses which provides a suitable habitat for the house dust mite. However, increasing concentrations of air pollutant gases, including NO_x, O_3 and others, may also be partly responsible (see sections 5.1 and 5.2).

References

Adelman, G. (ed.) (1987) *Encyclopedia of Neuroscience*, Vol. 2, Birkhauser, Boston.

Advances in Chemistry Series (1959) 21: Ozone chemistry and technology. American Chemical Society, Washington, DC.

Alloway, B. J. (1990) Chapter 3 in Alloway, B. J. (ed) *Heavy Metals in Soils*, Blackie and Son, Glasgow.

American Conference of Government Industrial Hygienists (1990) *Documentation of the TLV and Exposure Indices*, (5th edition), Cincinnati.

Ballunas, S. and Jastrow, R. (1990) Evidence for long-term brightness changes of solar-type stars. *Nature*, **348**, 520–522.

Battarbee, R. W., Flower, R. J., Stevenson, A. C. and Rippey, B. (1985) Lake acidification in Galloway: a paleoecological test of competing hypotheses. *Nature*, **314**, 350–352.

Bowen, H. J. M. (1979) *The Environmental Chemistry of the Elements*, Academic Press, London.

Callender, G. S. (1958) On the amount of CO_2 in the atmosphere. *Tellus*, **10**, 243–248.

Canada Council of Ministers of the Environment Interim Canadian Environmental Quality Criteria for Contaminated Sites. Report CCME EPC-C534, Winnipeg, Manitoba.

Central Statistical Office (1991) *Economic Trends*, HMSO, London.

Close, F. (1990) *Too Hot to Handle*, Allen, London.

Collier, J. G. and Hewitt, G. F. (1987) *Introduction to Nuclear Power*, Hemisphere, Washington DC.

Crouch, D. (1986) *Sci. of the Total Environ.*, **53**, 201–216.

Daily Telegraph (1991) The Citela battery car (14 November).

Davies, D. J. A. and Bennett, B. G. (1983) *Exposure Committment Assessments of Environmental Pollutants*, Vol 3 Monitoring and Assessment Research Centre, London.

Department of the Environment (1987) *Interdepartmental Committee for the Reclamation of Contaminated Land List of Trigger Concentrations for Contaminants*, DOE, London.

Dickinson, R. E. and Cicerone, R. J. (1986) Future global warming from atmospheric trace gases. *Nature*, **319**, 109–115.

Dunne, J. M. (1990) Vehicle emission control technology, in Dunderdale, J. (ed.), *Energy and the Environment*, Royal Society of Chemistry, Cambridge, p.208.

Epstein, J. M. and Gupta, Raj (1990) *Controlling the Greenhouse Effect*, Brookings Institute, Washington DC.

Fergusson, J. E. (1982) *Inorganic Chemistry and the Earth*, Pergamon Press, Oxford.

Fergusson, J. E. (1990) *The Heavy Elements: chemistry, environmental impact and health effects*, Pergamon Press, Oxford.

Gardner, A. F., Gillett, R. S. and Phillips, P. S. (1992) The menace under the floorboards. *Chem. in Brit.*, 344–348.

Greenwood, N. N. and Earnshaw, A. (1984) *Chemistry of the Elements*, Pergamon, Oxford.

Grubb, M. (1991) *Energy Policies and the Greenhouse Effect: Future Trends, Population and Development*, Vol. 1, Dartmouth Publications, Aldershot.

Hall, A. (1986) *Nuclear Politics: The History of Nuclear Power in Britain*, Penguin, Harmondsworth.

Harte, J., Holden, C., Schnieder, R. and Shirley, C. (1991) *Toxics A to Z*, University of California Press, Berkeley and Los Angeles.

Houghton, J. T., Jenkins, G. J. and Ephraims, J. J. (1990) *Climate Change: the IPCC Assessment*, Cambridge University Press, Cambridge.

Innes, J. L. (1987) *Air Pollution and Forestry*, HMSO, London.

Innes, J. L. (1990) *Assessment of Tree Condition*, HMSO, London.

Jackson, D. (1992) in *The Chemical Industry — Friend to the Environment*, Drake, J. A. G. (ed.), Royal Society of Chemistry, Cambridge.

Janes, S. M. and Cooke, R. U. (1987) Stone weathering in S.E. England. *Atmos. Environ.*, **21**, 1601–1622.

Johnston, H. S. (1975) Ground level effects of supersonic transport in the stratosphere. *Accounts Chem. Res.*, **8**, 289–294.

Kabata-Pendias, A. and Pendias H. (1984) *Trace Elements in Soils and Plants*, CRC Press, Boca Raton, Fl.

Kazantis, G. (1980) Chapter 8 in Waldron, H. A. (ed) *Metals in the Environment*, Academic Press, London.

Kloke, A., Sauerbeck, D. R. and Vetter, H. (1984) in Nriagu, J. O. (ed) *Changing Metal Cycles and Human Health*, Springer-Verlag, Berlin.

Krauskopf, K. B. (1967) *Introduction to Geochemistry*, McGraw-Hill, New York.

Lamarsh, J. R. (1983) *Introduction to Nuclear Engineering* (2nd edition), Addison-Wesley, Reading, Mass.

Langard, S. (1980) Chapter 4 in Waldron, H. A. (ed) *Metals in the Environment*, Academic Press, London.

Leclercq, I. (1988) *On the Track of the Scourge of the Forests*, DEFORPA, Paris.

Manahan, S. E. (1991) *Environmental Chemistry* (5th edition), Lewis Publishers, Chelsea, Mich.

Marples, D. R. (1986) *Chernobyl and Nuclear Power in the USSR*, MacMillan, Basingstoke.

Mattson, E. (1990) Paper 16 in *Cu '90: Refining, Fabrication, Markets*. Proceedings of Conference in Vasteras, Sweden, October, Institute of Metals, London.

Medvedev, Z. A. (1979) *Nuclear Disaster in the Urals*, Angus & Robertson, London.

Mellor, J. W. (1923) *Comprehensive Treatise on Inorganic and Theoretical Chemistry*, Longmans, London, **4**, 53–154.

Mellor, J. W. (1932) History of uranium. *Comprehensive Treatise on Inorganic and Theoretical Chemistry*, Longmans, London, **12**, 1–14.

Ministry of Housing, Physical Planning and Environment, Directorate General for Environmental Protection (Netherlands) (1991) *Environmental Standards for Soil and Water, 1991*. Leidschendam.

Moen, J. E. T., Cornet, J. P. and Evers, C. W. A. (1986) in Assink, J. W. and van den Brink, W. J. (eds) *Contaminated Soil*, Martinus Nijhoff, Dordrecht.

Molina, M. J. and Rowland, F. S. (1974) Stratospheric sink for chlorofluoromethanes, chlorine atom catalysed destruction of ozone. *Nature*, **249**, 810.

Murley, L. (ed) (1992) *Pollution Handbook*, National Society for Clean Air and Environmental Protection, Brighton.

National Academic Press (1981) *Indoor Pollutants*, Washington DC.

Nero, A. V. (1979) *A Guidebook to Nuclear Reactors*, University of California, Berkeley.

Newman, P. and Kenworthy, J. (1989) *Cities and Automobile Dependence*, Gower, Aldershot.

Nriagu, J. O. (1988) *Environ. Pollut.*, **50**, 139–161.

Nuclear Energy Agency (1986) *Projected Costs of Generating Electricity from Nuclear and Coal Fired Stations Commissioning 1995*, OECD, Paris.

Nuclear Energy Agency (1989) *Nuclear Accidents: Intervention Levels for Protection of the Public*, OECD, Paris.

Nuclear News (1992) *World List of Nuclear Power Plants*, **35**, 55–74.

O'Neill, P. (1990) Chapter 5 in Alloway, B. J. (ed) *Heavy Metals in Soils*, Blackie and Son, Glasgow.

Openshaw, S. (1986) *Nuclear Power Siting and Safety*, Routledge & Kegan Paul, London.

Owen, K. (1981) Technology: How the "Bedford smell" foxed the experts, The Times (London), January 23.

Packham, R. F. (1990) Chapter 5 in Harrison, R. M. (ed) *Pollution: Causes, Effects and Control* (2nd edition) Royal Society of Chemistry, London.

Pacyna, J. M. (1987) in Hutchinson, T. C. and Meema, K. M. (eds) *Lead, Mercury, Cadmium and Arsenic in the Environment*, SCOPE 31, John Wiley and Sons, Chichester.

Pahl, D. A., Zimmerman, D. and Ryan, R. (1990) An overview of combustion emissions in the USA in Clement, R. and Kagel, R. (eds), Lewis, Boca Raton.

Park, C. (1987) *Acid Rain*, Routledge, London.

Patterson, W. C. (1976) *Nuclear Power*, Pelican, Harmondsworth.

Patterson, W. C. (1985) *Going Critical*, Paladin, London.

Peters, W. C. (1978) *Exploration and Mining Geology*, John Wiley and Son, New York.

Pocock, R. F. (1977) *Nuclear Power: Its Development in the UK*, Unwin, Old Woking.

Roberts, A. and Medvedev, Z. (1977) *Hazards of Nuclear Power*, Spokesman.

Roberts, L. E. J., Liss, P. S. and Saunders, P. A. H. (1990) *Power Generation and the Environment*, Oxford University Press, Oxford

Rose, A. W., Hawkes, H. E. and Webb, J. S. (1979) *Geochemistry in Mineral Exploration* (2nd edition) Academic Press, London.

Rotty, R. M. (1983) Annual emissions of CO_2 by fuel type. *J. Geophys. Res.*, **88**, 1301.

Rowland, A. J. and P. Cooper (1983) *Environment and Health*, Edward Arnold Ltd, London.

Schroeter, H. (1985) *Report of the Acid Rain Enquiry*, Scottish Wildlife Trust, Edinburgh.

Social Trends (1991), HMSO, London.

Speakman, K. (1990) Emission control — statutory requirements, in Dunderdale, J. (ed.) *Energy and the Environment*, Royal Society of Chemistry, Cambridge.

Spence, R. D., Rykiel, E. J. and Sharpe, P. J. H. (1990) Ozone alters carbon allocation in loblolly pine: assessment with carbon-11 labelling. *Environ. Pollution*, **64**, 93–106.

Sprague, A. (1990) *Statistical Review of World Energy*, British Petroleum, London.

Stephens, N. (1980) *Three Mile Island*, Junction.

Steinnes, E. (1990) Chapter 11 in Alloway, B. J. (ed.) *Heavy Metals in Soils*, Blackie, Glasgow.

Tepper, L. B. (1980) in Waldron, H.A. *Metals in the Environment*, Academic Press, London.
Tolba, M. K. (1983) *Water Quality Bull.*, **8**, 115–120, 167.
Ulrich, B. (1980) *Soil Science*, **130**, 193–199.
Ulrich, B. (1990) Waldsterben: forest decline in West Germany. *Environ. Sci. Technol.*, **24**, 439.
Waldron, H. A. (1980) Chapter 6 in Waldron, H. A. (ed) *Metals in the Environment*, Academic Press, London.
Wayne, R. (1991) *Chemistry of Atmospheres*. Oxford University Press, Oxford.
Wilson, B. J. (ed.) (1986) *The Radiochemical Manual*, Radiochemical Centre, Amersham.
World Health Organisation (WHO) (1984) *Guidelines for Drinking Water Quality Vol 1, Recommendations*, WHO, Geneva.
Wright, R. F. (1976) *Ambio*, **5**, 219.

Further reading   *Ozone*

Banks, R. E. (ed.) (1979) *Organochlorine Chemicals and Their Industrial Applications*, Ellis Horwood, Chichester.
Knunyants, I. L. and Yakobson, G. G. (eds) *Organofluorine Chemicals and Their Industrial Application*, Ellis Horwood, Chichester.
Nebel, C. (1981) Ozone in *Encyclopedia of Chemical Technology (3rd edition)*, Wiley, New York.

Oxides of carbon, nitrogen and sulphur

Tyler Miller, G. (1990) Living in the Environment (6th edition), Wadsworth, London.
Flavin, C. (1989) *Slowing global warming, a worldwide strategy, Worldwatch Paper*, **91**, Washington DC.
Grainger, A. (1990) *The Threatening Desert*, Earthscan.
Meetham, A. R. (1981) *Atmospheric Pollution* (4th edition), Pergamon, Oxford (appendix on units).
Warrick, R. A., Barrow, E. M. and Wigley, T. M. L. (1990) *The Greenhouse Effect and its Implications for the European Community*, EC, Luxembourg.
Brimblecombe, P. (1987) *The Big Smoke*, Methuen, London.

Heavy metals

Alloway, B. J. (ed) (1990) *Heavy Metals in Soils*, Blackie and Son
Bowen, H. J. M. (1979) *The Environmental Chemistry of the Elements*, Academic Press, London.
Fergusson, J. E. (1990) *The Heavy Elements: Chemistry, Environmental Impact and Health Effects*, Pergamon Press, Oxford.
Waldron, H. A. (ed) (1980) *Metals in the Environment*, Academic Press, London.

Radionuclides

Allaby, M. (ed.) (1988) *Dictionary of the Environment*, MacMillan, Basingstoke.
Burn, D. (1978) *Nuclear Power and the Energy Crisis*, MacMillan, Basingstoke.
Eisenbud, M. (1987) *Environmental Radioactivity from Natural, Industrial and Military Sources* (3rd edition), Academic Press, Orlando.
Hines, L. G. (1988) *The Market, Energy and Environment*, Allyn & Bacon, Boston.
Hoyle, F. (1977) *Energy or Extinction*, Heineman.

Judd, A. M. (1991) *Fast Breeder Reactors, an Engineering Introduction*, Pergamon, Oxford.

Lilienthal, D. (1980) *Atomic Energy: a New Start*, Harper and Row.

Marshall, W. (ed.) (1983) *Nuclear Power Technology*, Vol. 3, *Nuclear Radiation*, Clarendon Press, Oxford.

Nuclear Installations Inspectorate (1992) *Safety Assessment Principles for Nuclear Plants*, NII, London.

OECD (1987) Chenobyl and the Safety of Nuclear Reaction in OECD Countries, OECD.

Parliamentary Select Committee on Energy (1990) 4th Report, *The Cost of Nuclear Power*, HMSO, London.

UK Royal Commission (1976) *6th Report. Nuclear Power and the Environment*, HMSO, London.

US Department of Energy (1990) *Radon Research Programme*, March.

Weart, S. R. (1988) *Nuclear Fear, A History of Images*, Harvard University Press, Cambridge, Mass.

Yearbook of Science and Technology (1992) McGraw-Hill, p.297

6 Organic pollutants

6.1
Smoke Smoke is a term applied to particulate carbonaceous particles of < 10 μm in diameter produced by the partial combustion of organic substances. Smoke rises upwards into the air above the fire due to its thermal buoyancy and its colour depends on the material being burnt. The amount of smoke is related to the combustion conditions, including the physical form of the fuel (droplets of oil, coal pieces or powder), the degree of mixing of volatile materials released on heating with air, and the temperature. The products of the complete combustion of an organic substance are CO_2, water vapour and ash (non-combustible inorganic residue). The formation of smoke and soot particles only occurs when the combustion is incomplete.

In the flame, hydrocarbon molecules are progressively broken down to smaller fragments and combine with oxygen to form CO. In the presence of adequate O_2, the CO becomes oxidized to CO_2. However, the formation of CO is relatively fast, but its oxidation to CO_2 is slower, and so if there is an inadequate supply of O_2, CO will accumulate. Soot formation generally accompanies the formation of CO and is due to an inadequate supply of air. Partially degraded fuel molecules can polymerize to produce carbon nuclei which then accumulate further material by surface adsorption and coagulation. The soot particles formed are usually < 1 μm in diameter and are similar in size to the wavelength of light so they can be effective in absorbing light and scattering it (Clarke, 1992).

Spray-atomized light fuel oils normally evaporate completely and produce little soot and smoke. With heavy fuel oils only the light fractions evaporate at first leaving a tarry residue which is gradually cracked (releases lower molecular weight volatile compounds) by the heat of the flame. However, if the temperature is not high enough, some particles < 50–100 μm in diameter (called smuts) are likely to be emitted from the chimney. If the oil has a high S content the carbon smuts can contain H_2SO_4 which makes them very corrosive to fabrics, paintwork and many other surfaces (Clarke, 1992).

Coal combustion proceeds in a similar way to oil, the volatile compounds burn as gases and the remaining material burns slowly leaving an ash residue. If the coal is pulverized first (as in coal-fired electricity generating stations), the combustion is more rapid and controllable and does not normally produce smoke. In contrast, an open fire in a grate with lumps of coal, a log fire, or a primitive cooking stove tend to form soot and smoke due to the inefficient mixing of air and the volatile compounds. Following the introduction of the Clean Air Act of 1956, smoke control was achieved by using smoke-free fuels in place of bituminous coal or wood (Clarke, 1992) (see sections 5.2 and 6.2).

In general, low temperature fires such as coal or wood fires in a grate, or uncontrolled fires such as bonfires, forest fires and burning cereal straw stubbles tend to produce more smoke than controlled hot fires in stoves or boilers. The importance of the material itself is shown by a comparison between diesel engine smoke which is three times darker than bituminous coal smoke and seven times blacker than petrol engine exhaust smoke (Elsom, 1987). Smoke is generally less of a problem in spark ignition petrol engines because the fuel and air is more thoroughly mixed than in diesel engines, especially when the latter are under a high torque load where there is a rich supply of fuel. The diesel smoke odours include partly oxygenated hydrocarbons such as aldehydes.

It was estimated that the disposal of straw by burning in open fields in the UK in 1984, produced 18 000 t of black smoke over a short period of two or three weeks in the main cereal growing areas (Figure 6.1) (RCEP, 1985). This significant source of smoke pollution continued annually until it was banned from 1993 for nuisance and environmental reasons.

Smog is a mixture of smoke and fog but where the smog is generated from coal combustion, SO_2 is frequently present also. The notorious London smog of December 1952 caused the death of 4700 people and

Figure 6.1 Burning cereal straw and stubble.

many thousands of others to be very ill with respiratory problems. The peak daily concentrations of atmospheric pollutants reached 6000 $\mu g/m^3$ smoke and 4000 $\mu g/m^3$ SO_2 (see section 5.2). In addition to these two characteristic pollutants, other toxic substances such as HF from industrial sources may also be present in high concentrations in urban and industrial areas and they can exacerbate the health effects of smoke and SO_2 (Elsom, 1987).

People suffering from respiratory allergies, such as asthma and hay fever, and people carrying out strenuous exercise are more susceptible to the effects of SO_2 and smoke than other people. This is partly because they usually need to breath through their mouths instead of their nostrils, and so there is more chance of the acid droplets and smoke going straight down the trachea without any filtering in the nostrils (Elsom, 1987).

The WHO short-term maximum concentrations of both SO_2 and smoke are 250 $\mu g/m^3$ (at which the condition of patients with respiratory illness would be expected to deteriorate) and 100 $\mu g/m^3$ for long-term exposure. At 500 $\mu g/m^3$ of smoke and SO_2, excess mortality can be expected among elderly and chronically sick people (Elsom, 1987).

Soot and smoke particles can have numerous organic pollutants adsorbed to their surfaces and most important among these are PAHs (see section 6.2). Some PAHs, such as benzo-[a]-pyrene, are carcinogenic and are probably the reason for the higher incidence of scrotal cancer in chimney sweeps which was observed almost 200 years ago (in 1775) by Percival Potts, a London surgeon (Rodricks, 1992). Tobacco smoke contains several hazardous substances including nicotine (a supertoxin), tar (with PAHs), formaldehyde, NO_x and CO (Murley, 1992). There is a vast body of evidence linking smoking with the incidence of cancer of the lung and certain other tissues. It has also been suggested that the higher concentrations of PAHs in urban air compared with rural areas are responsible for a higher incidence of lung cancer in urban residents. Fortunately, however, the introduction of clean air policies in both the UK and the USA have resulted in marked decreases in PAHs in air, with mean annual benzo-[a]-pyrene concentrations in air decreasing in London by a factor of ten between 1935 and 1965 and by a factor of 3 between 1966 and 1975 in the USA.

As a result of the controls brought in after the Clean Air Act of 1956 in the UK, smoke emissions decreased by 85% and SO_2 emission by 40% between 1956 and the late 1980s (Timberlake and Thomas, 1990).

The most recent EC Directive on smoke emissions (80/779/EEC) specifiied:

1 hour smoke   = 80 $\mu g/m^3$
winter smoke   = 130 $\mu g/m^3$
peak smoke   = 250 $\mu g/m^3$

Smoke is also a serious indoor pollution problem (see chapter 2 and section 6.3). Although the WHO guideline value for particulates (smoke) in indoor air is 100–150 µg/m, rural dwellings in Third World countries frequently exceed these values and concentrations of < 14 000 µg/m have been recorded. These high levels of smoke are due to burning wood, animal dung or coal in inefficient stoves and poor ventilation restricting smoke escape. Respiratory diseases in both children and adults are linked to indoor air pollution of this type.

In developed countries, indoor pollution can include tobacco smoke and many of the pollutants listed in Table 2.5. With increased awareness of the need for energy conservation, houses are generally less well ventilated nowadays due to the exclusion of draughts and so indoor pollutants are more likely to accumulate, although some do undergo precipitation or transformation within the house (see section 2.4.3).

**6.2
Methane and other
hydrocarbons – coal
and oil as sources**

6.2.1   The formation of coal

Coal is formed by the compaction and metamorphosis of the residues of woody plants ultimately under high temperatures and the pressure of overlying sediments accumulated over 50 million years or longer. The initial components are cellulose and lignin. *Cellulose* is a linear polymer (**1**) of glucose with a molecular weight in the range $40 \times 10^3 - 150 \times 10^3$ and typical composition $(C_6H_{10}O_5)_n$. *Lignin* is a three-dimensional polymer

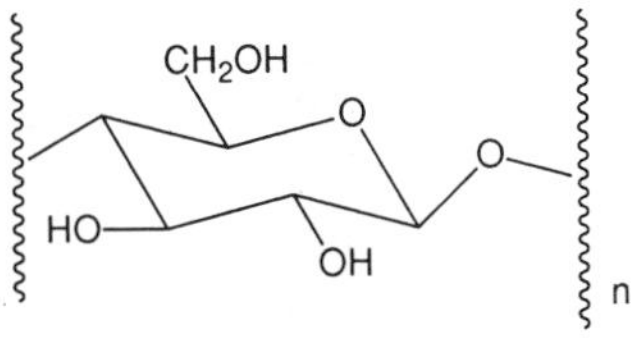

1

interwoven with cellulose and constructed from units (**2**) derived from cinnamic acid. The growth of the polymer is achieved largely through oxygen atoms in ether linkages. One typical section of lignin is shown as the dimer unit in structure (**2**).

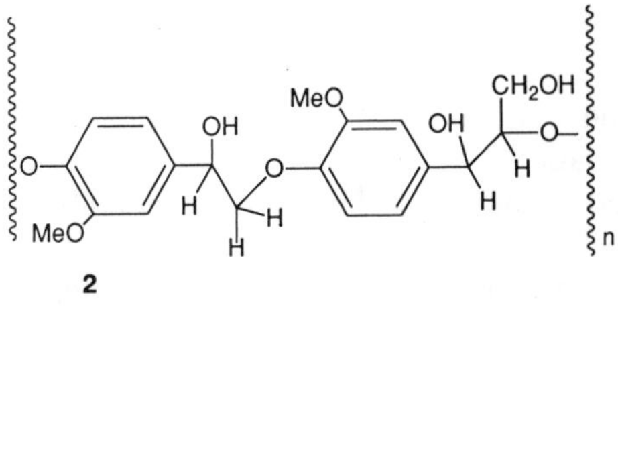

2

Table 6.1   Changes in the composition of coal substances as they mature

	Typical percentage composition				
	C	H	O	N	S
Wood	56	6.5	37.5	—	—
Peat (humus)	55	6	30	1	1.3
Lignite	73	4	21	1.5	0.5
Anthracite	93	2.5	2.5	1	0.5

A representative composition of lignin is $(C_{10}H_{12}O_4)_n$. Anaerobic bacteria depend for their oxygen on that combined in these plant residues and hence there is a progressive reduction in combined oxygen as coalification proceeds (Table 6.1).

Lignite is the lowest ranking coal and anthracite the highest. The majority lie between these two extremes and contain 85–90% carbon.

6.2.2   Petroleum

Historical.   Petroleum has been known in years BC to occur in surface seepages and was first obtained in pre-Christian times by the Chinese. The modern industry had its beginnings in Romania and in wells sunk by Colonel E. A. Drake in Pennsylvania in 1859. The principal early use was to replace expensive whale oil for lighting, but today its consumption as a fuel and its dominance of the market for chemicals has led since 1965 to a doubling of proven reserves to 140 Gt. In the UK about 70 Mt is used for fuel and 20 Mt for export and as chemical feedstock. The daily worldwide consumption is now about 65 million barrels or 3 Gt a year.

Formation.   Petroleum is largely formed biogenetically from matter deposited in shallow seas and subsequently compressed by the overburden of deposited clays and shales. An intermediate coal-like material formed by bacterial action on the deposits is known as kerogen. This may be one of three types formed from (a) algae, (b) marine plankton or (c) higher plants. Hydrogen sulphide, a typical product of organic decay, may also be formed.

Properties.   The major compounds are saturated alkanes; alkenes are absent but aromatic hydrocarbons may occur to a significant extent, for example in Borneo deposits. The environmentally significant polycyclic aromatic hydrocarbons are discussed at the end of this chapter (p. 206). Other organic components of petroleum, including some which contain

Aliphatic compounds include the volatiles:

Methane CH$_4$ through the range of straight and branched chains
Ethane   C$_2$H$_6$ up to C$_{76}$H$_{154}$
Propane C$_3$H$_8$

Alicyclic and aromatic components include:

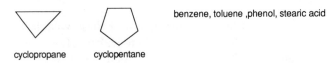

benzene, toluene ,phenol, stearic acid

cyclopropane   cyclopentane

Nitrogen compounds are typified by:

Sulphur compounds include:

thiols R-SH Thioethers R-S-R ; Ph-S-Ph and aromatics including:

thiophenol   a phenyl alkyl-   thiophene   benzothiophene   dibenzothiophene
               thioether

Figure 6.2   Some organic components of petroleum.

Table 6.2   Boiling points of typical components of petroleum (°C)

Methane	−161	Benzene	80
Propane	−42	Toluene	111
n-Butane	−0.5	*p*-Xylene	138
n-Hexane	69	Naphthalene	218 (m.p.80)
n-Octane	126	Anthracene	340 (m.p.216)

nitrogen and sulphur, are shown in Figure 6.2. Table 6.2 lists the boiling points of some typical substances.

Petroleum deposits are complex mixtures which may include as many as 40 individual substances which are not economically separable by distillation. Hence commercial practice is to sub-divide within a given boiling range (°C):

Natural gas: methane, propane, butanes

Light naphtha: 20–100
Heavy naphtha: 100–150
Kerosene: 150–235
Light gas oil: 235–345
Heavy gas oil: 345–565 (steam assisted)

Release to the environment.   Clearly manipulation of these volatile fractions on the present massive scale leads inevitably to leakage. Losses in the
oil industry are estimated at 6.5% of the total of about 10 Mt/year, while
major subsequent losses are attributed to cars and HGVs (41%) and to the
industrial use of solvents (40%). Natural gas and minor sources such as
domestic, incineration and power stations account for 12% of the remainder. The rapid rise in the bulk of hydrocarbon processing is shown by the
development of the petrochemicals industry in the UK from its beginnings
in the 1920s to its eight-fold growth between 1940–1955, when 42 Mt were
processed (cf. p. 223). In 1986 the North Sea field produced 177 Mt of oil
and 96×10^9 m^3 of natural gas and met 20% of Western Europe's energy
needs, equivalent to the 400 Mt of coal mined at that time (Martin, 1990).
These figures may be compared with those given for the USA (British
Petroleum, 1991) which produced 417 Mt of oil and 444 Mt coal equivalent
of natural gas in 1990.

6.2.3   Methane

Owing to its importance as a greenhouse gas (Table 5.3) methane is best
considered separately. It occurs at 1.7 ppm in air and in addition to losses
from processing natural gas it arises from the following biological sources:

a) cultivation of rice;
b) anaerobic fermentation in swamps and rain forests;
c) in the intestines of animals, chiefly cattle.

Its effect on surface warming is enhanced by a factor of seven compared to
carbon dioxide because it absorbs infrared radiation near 3000 cm^{-1}, which
coincides with a window in the spectrum of carbon dioxide (Figure 5.5).

Reactions in air.   The hydroxyl radical features critically in the chemistry
of air pollution; it is formed when ozone is photolysed (equation 5.5) to
afford excited oxygen which combines with water vapour (Wayne, 1991):

$$O^{\cdot} + H_2O \rightarrow HO^{\cdot} + .OH \qquad (6.1)$$

The ˙OH radical reacts with CO in clean air to form CO_2 (equation 6.2)
and also abstracts H˙ atom from methane in polluted air (equation 6.3) so
converting it into the reactive methyl radical:

$$HO^{\cdot} + CO \rightarrow H^{\cdot} + CO_2 \qquad (6.2)$$
$$HO^{\cdot} + CH_4 \rightarrow {}^{\cdot}CH_3 + H_2O \qquad (6.3)$$

This in turn captures oxygen to give the methylperoxy radical. The natural outcome is the interaction between peroxy radicals to regenerate oxygen, but NO in polluted air diverts the system according to equation 6.4:

$$CH_3O\text{-}O^{\cdot} + NO \rightarrow CH_3O^{\cdot} + NO_2 \qquad (6.4)$$

An important general point now follows, for any alkoxy radical of this type may be converted into the corresponding carbonyl compound by donating an H atom to any acceptor molecule (M). The mechanism of the electron pairing is shown for the methoxy radical in equation 6.5:

$$(6.5)$$

6.2.4 Higher alkanes

These reactions are of the same type, for example ethane is converted into acetaldehyde which, like formaldehyde, is a volatile lachrymatory substance which also contributes to smog. The further reactions of acetaldehyde are particularly significant since they lead to the formation of peroxyacetylnitrate (PAN), one of the principal irritants in photochemical smog. In the first step (equation 6.6) a relatively stable acetyl radical is formed from acetaldehyde by abstraction of an H atom:

$$Me\text{-}\overset{\bullet}{C}{=}O + H_2O \qquad (6.6)$$

After capture of oxygen, which is available in high concentration, the acetylperoxy radical is obtained:

$$Me\text{-}\overset{\bullet}{C}{=}O + O_2 \longrightarrow \qquad (6.7)$$

Notice that NO_2 is itself an odd electron species and pairs to give PAN:

$$(6.8)$$

PAN is formed naturally by photooxidation of natural plant terpenes but at levels well below 1 ppb. In European cities the levels reach 10 ppb and in extremely polluted areas these may reach 50 ppb (Harrison, 1992). PAN is highly phototoxic and contributes to leaf damage (discussed in the section on acid rain, p. 134).

The formation of similar alkoxyperacetylnitrates will also take place with other volatile hydrocarbon starting materials. PAN may itself be generated following fragmentation of higher alkyl radicals obtained by the action of the same OH initiator:

$$\text{PAN} \longleftarrow \quad \underset{Me}{\overset{\backslash}{/}}{=}O + Me^{\bullet} \qquad (6.9)$$

A diurnal urban pollution cycle. Figure 6.3 shows a typical 24 hour experience.

In unpolluted air O_2 and NO_x achieve a steady tropospheric state including a little ozone (p. 114) but the injection of CO and unburnt hydrocarbons into the atmosphere diverts this equilibrium:

$$\qquad (6.10)$$

This can be explained by reference to equation 6.4 since in this sequence NO_2 is produced via R–H *without participation of ozone.* Entry of NO_2 from this source gives additional [O] which combines with oxygen to generate abnormally high levels of ozone together with aldehydes (cf. equation 6.5).

Reference to Figure 6.3 shows the typical rise in airborne hydrocarbons in the morning as traffic density rises to reach a peak near 08.00 h. Meanwhile the strengthening sun induces the reactions leading to aldehydes and ozone as explained above. With traffic growth the concentration of NO is also seen to rise and it subsequently reacts to form NO_2. By mid-morning appreciable amounts of oxidized hydrocarbons and NO_2 are

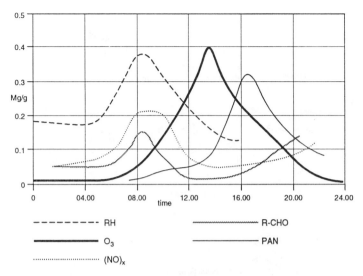

Figure 6.3 Cyclic concentration changes of pollutants.

available and given strong sunlight formation of PAN (equations 6.6–6.8) proceeds apace to reach a peak in the afternoon.

The hydroxyl radical and aromatic hydrocarbons. Benzene (p. 216) is itself a significant air pollutant (WHO, 1987) especially as it is now used to formulate lead-free petrols. Owing to the stability of the aromatic system it reacts principally by hydrogen atom abstraction (cf. equation 6:3) with ring retention and with the formation of phenols:

$$\text{benzene} \xrightarrow{\overset{H}{\cdot}\text{OH}} \text{phenyl radical} \xrightarrow{O_2} \text{Ph}-O-\overset{\cdot}{O} \xrightarrow[NO_2]{NO} \text{Ph-}\overset{\cdot}{O} \xrightarrow{R\text{-}H} \begin{array}{c}\text{Ph-OH}\\+R^{\cdot}\end{array} \quad (6.11)$$

The entry of phenols in this way into the environment has serious consequences as they are acidic and often toxic.

The side chains of benzene homologues may react in a similar fashion to methane especially as abstraction of hydrogen gives the stabilized benzyl radical. This will capture oxygen (equation 6.12) and then undergo analogous changes to those shown in equations 6.4 and 6.5:

$$\overset{CH_3}{\bigcirc} \xrightarrow{\overset{\cdot}{O}H} \underset{\text{benzyl radical}}{\overset{\overset{\cdot}{C}H_2}{\bigcirc}} \xrightarrow{O_2 \text{ etc}} \underset{\text{benzaldehyde}}{\overset{\overset{H}{C}=O}{\bigcirc}} \quad (6.12)$$

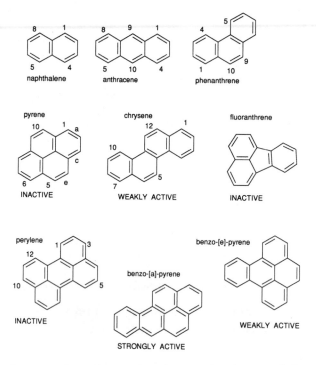

Figure 6.4 Structures of some benzene homologues and environmentally important PAH.

6.2.5 *Polycyclic aromatic hydrocarbons (PAH)*

The initial fusion of two benzene rings can give rise to only one product – naphthalene – but attachment of a third ring can give the linear homologue anthracene, or the branched isomer phenanthrene (Figure 6.4). These substances occur in considerable amounts in natural fuel deposits and are formed during coking of coal and on pyrolysis of fossil fuels. They are not included in the definition of a PAH which is reserved for higher homologues containing four or more benzene rings.

The number of individual PAH is considerable, bearing in mind the permutations of the modes of ring fusion and also the various positions available for insertion of side chains. As a consequence Figure 6.4 shows only a limited selection which includes those of commonest occurrence and/or highest toxicity.

Historical note. Industrial disease due to contact with PAH was first identified in chimney sweeps by Percival Pott in 1775. These children normally worked from age eight into adulthood, although 'apprenticeships' could start at age four. A typical consequence of this cruel practice was the

Table 6.3 Mortality from cancer in gas workers in Great Britain

	Increase in deaths/100 000	
	Gas workers	National rate
Cancer of the lung	3.82	2.13
Cancer of the bladder	0.40	0.17
Cancer of the skin and scrotum	0.12	0.02

development of skin cancers probably induced by benzo-[a]-pyrene (Figure 6.4) which occurs at 0.2% in soot. In 1875 von Volkman showed that there was a link to skin cancer of workers in the German coal tar industry.

Later notable contributions were the link to lung cancer in workers in the coal tar and gas industries which was made by the Kennaways in 1947. This was extended in 1972 by Sir Richard Doll who had made a study of mortality in a large group of workers in the same industry over a 12-year period (Table 6.3). The history of the evaluation of PAH carcinogens has been reviewed by Sir Ernest Kennaway (1955).

Toxicity.   The level of activity is given alongside the structures in Figure 6.4. Benzo-[a]-pyrene (BaP) is regarded as the most dangerous of the group as it is widely distributed and strongly carcinogenic. In a test on the skin of mice, activity was noted at a dose level of 5.6×10^{-5} mmol; liver damage and teratogenicity were also observed in mice. In rats 10 monthly intra-tracheal injections of 0.05 mg produced lung tumours in 28% of test animals (Osborne, 1987).

Four fused benzene rings are a necessary but not sufficient condition for activity and a subtle structural factor is revealed since benzo-[e]-pyrene is much less active than its [a] isomer. It is known that initiation requires the binding of the PAH to cellular macromolecules DNA, RNA or protein. A key step in the activation of benzo-[a]-pyrene is its conversion into the epoxide (structure (3), see p. 41) and that the ease of oxidation of PAHs by free radicals is indicative of their level of activity. Oxidized derivatives of PAHs also occur in the environment.

3

Table 6.4   Worldwide production of PAH (10^3tons/yr)

Source	Quantity	%
Heating and power production	260	51
Industrial producers	105	20
Incineration and open burning	135	28
Vehicular transport	4.5	0.9

Table 6.5   Principal PAHs in emitted particles (ng/mg)

Pyrene	18–105	Benzo-[a]-pyrene	1–7
Fluoranthrene	45–203	1-Nitropyrene	6–10

A full discussion of testing procedures and structure–activity factors is given by Harvey (Harvey, 1985) and by the International Agency for Research on Cancer (1983).

Sources and distribution.   The production of PAH in 1976 was as shown in Table 6.4 (Sues, 1976). In the early 1980s BaP emissions in the USA were of the order of 10^3 tons/yr from these sources and individuals were likely to inhale up to 1.5 μg/day. The recommended acceptable level is 0.2 μg/m^3 which was always exceeded for coke oven workers. After 5 years of exposure their risk of infection was increased ten-fold.

PAH are found at low levels of ng/g in mineral oils and in paraffin wax. They are formed by pyrolysis in petrol and light diesel engines owing to the limited air supply. If a catalytic converter (p. 128) is not fitted BaP is emitted at levels up to 50 μg/km travelled; the converter reduces this to a range of 0.05–0.3 μg/km. Aircraft engines typically emit 10 mg of BaP during each minute of operation. The principal components of particles are given in Table 6.5 (IARC, 1983). The detection of the nitropyrene is of interest and a number of other nitroderivatives have been isolated. They clearly originate from contact with nitrogen fixation products within the engine (p. 126).

PAHs enter the water body from fall-out from the above sources and also from run-off from bitumen treated roadways. The worldwide use of this material is estimated at 60×10^6 tonnes/yr with 2×10^6 tonnes applied annually in the UK and ten times this amount in the USA. Some levels in water bodies are given in Table 6.6 (IARC, 1983). The levels in domestic sewage vary but rise markedly during periods of heavy rain, for example pyrene then rises to 16×10^3 and BaP to 1800 ng/l.

PAH are found in food either by deposition on leaves (lettuce) or from heated fats (steak, sausage) (Table 6.7) (IARC, 1983).

Within the house occupants are also exposed to risk from emissions from

Table 6.6 PAH levels in water (ng/l)

	Thames (Kew)	Factory effluent	Industrial sewage	Domestic sewage
Fluoranthrene	140	2200	3400	—
Pyrene	—	1960	3120	250
Benzo-[a]-pyrene	130	100	100	1

Table 6.7 Benzo-[a]-pyrene in foodstuffs (μg/kg)

Charcoal broiled steak	8	Dried flour	4
Margarine	1–36	Toast bread	0.5
Sausages	4–50	Lettuce	3–12
Roasted coffee	1–13		

Table 6.8 PAH in tobacco smoke (μg/100 cigarettes)

(Anthracene)	2–23	Benzo-[a]-pyrene	0.5–8
Chrysene	0.5–9	Pyrene	5–27
5-Me-chrysene	0.06	Carbazole (**4**)	100

heating, for example gas central heating in a non-smoking household generates BaP in the range 0.1–0.6 ng/m^3. For smokers these limits lie between 0.4 and 1.8 ng/m^3.

The risk to smokers is of special concern and a detailed analysis of the smoke has been made (IARC, 1986). 45 individual compounds were detected including some heterocyclic N-compounds such as carbazole (**4**) (Table 6.8). It is worth noting that a number of methylated PAH were detected including several other methylchrysenes at levels up to 7 ng/cigarette. The 5-Me isomer is the most toxic of these isomers.

4

Metabolism of aromatic compounds.   A principal route in the metabolism of aromatic hydrocarbons is the formation of arene oxides which then

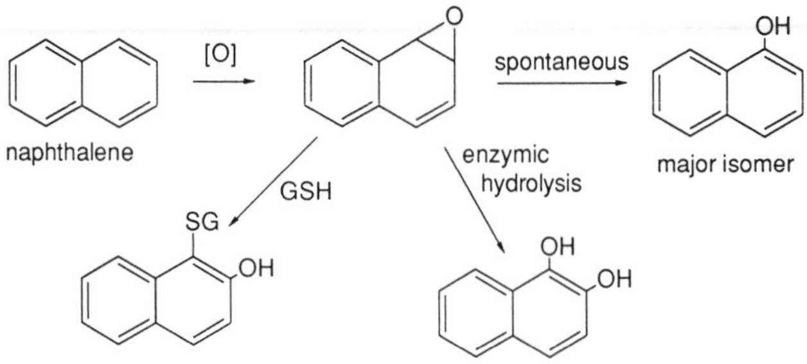

Figure 6.5 Oxidative metabolism of aromatic compounds.

isomerize to phenols (Figure 6.5) (Jerina, 1974). In mammals, including man, ingestion of these substances induces changes in the P-450 cytochrome enzymes within the endoplasmic reticulum. This is principally sited in the liver, but also in the kidneys, lungs and intestines. Oxides and phenols formed in this way become conjugated with water-solubilizing molecules and are then excreted. One such is the tripeptide glutathione–glutamyl cysteinyl glycine (**5**). Its combination (as GSH) is shown in Figure 6.5.

An alternative and competing hydroxylation path is illustrated for PCBs in section 6.4.

6.3
Organic solvents   Organic solvents are today produced almost entirely from petroleum (pp. 222–3) and units for their production and plant for their development in chemical synthesis are commonly sited in close proximity. The relative importance of individual compounds is indicated by production levels in the United States (1984) shown in Table 6.9, which includes those of the primary sources ethylene, propylene and 1,3-butadiene.

Table 6.10 lists those solvents in most common use with their principal properties and applications.

As will be appreciated, a detailed study of solvents encompasses a large proportion of organic chemistry and the discussion must necessarily be

Table 6.9 Production of organic chemicals in the USA (1000 tons)

Ethylene	13	Ethylene glycol	2.0
Propylene	6.3	p-Xylene	1.8
Toluene	6.1	Acetic acid	1.3
Benzene	5.5	Phenol	1.2
Styrene	3.1	Butadiene	1.0
Vinyl chloride	3.1	Acetone	0.8
Methanol	3.0	Cyclohexane	0.7

Table 6.10 Properties of some important solvents (Verschueren, 1983)

Compound	b.p.	Odour perception	Applications
Acetic acid	118	0.1–1.5 mg/m^3	Synthesis, pharmaceutical, photographic, polymers
Acetone	56	5–50 mg/m^3	Solvent for cellulosic resins and vulcanisates, adhesives, paints
Benzene	80	3 mg/m^3	Motor fuels, synthesis especially maleic anhydride, flavours, perfumes, paints
Carbon tetrachloride	77	500 mg/m^3	Manufacture of fluorocarbons
Chloroform	62	1000 mg/m^3	Refrigerants, pharmaceuticals
Cyclohexane	81	4–10 mg/m^3	Nylon synthesis
1,4-Dioxane	101	4–25 mg/m^3	Lacquers, paints, cosmetics, cleaning and detergency
Ethanol	78	5–50 mg/m^3	Synthesis, brewing, soaps, cosmetics
Ethyl ether	35	1–10 mg/m^3	Manufacture of ethylene, perfumery, extraction
Methanol	65	1–100 mg/m^3	Synthesis of proteins, esters and H.CHO for plastics. Dehydration of natural gas
Tetrahydrofuran	66	10 mg/m^3	Polyether synthesis, PVC solvent
Toluene	111	1–100 mg/m^3	Fuels, paints and coatings, isocyanates, benzene substitute
Trichloroethane	74	540 mg/m^3	Metal cleaning, protection of upholstery, adhesives
p-Xylene	138	1–10 mg/m^3	Synthesis of terephthalic acid, motor fuels

limited to those which offer the greatest environmental risk. The following brief case studies are therefore concerned with those which are especially toxic and those which are most readily released to air.

6.3.1 Adhesives

These are formulated with volatile solvents so that once applied they rapidly become tacky and ready to bond. The solvents are therefore liable to escape when used domestically (p. 221) and in small factories with inadequate control measures.

In 1983 the demand in the USA was 1.53 and in Europe 0.69 Mt with usage in Japan about the same as in Europe. This consumption grew at almost 4% until these outlets consumed about 3.36 Mt by 1988 (Table 6.11) (Montreal Protocol, 1991). 1,1,1–Trichloroethane (TCE) is one of the most used solvents as it is non-flammable and dries rapidly. Between 40–50 thousand tonnes are used annually for adhesive production. It is dangerous when inhaled by 'glue sniffers' and can cause death through congestion of the lungs or from heart failure (Glowa, 1990). Other solvents in common use are CH_2Cl_2, toluene, n-hexane, methyl ethyl ketone and ethyl acetate. Major particular uses are in bonding decorative laminates to

Table 6.11 Demand for adhesives by sector in USA and Europe

Application	Weight (10^6 tonnes)	%
Packaging	2500	42
Non-rigid bonding	1100	19
Construction	1000	17
Tapes	500	8.5
Rigid bonding	400	7
Transportation	300	5
Domestic market	100	1.5

particle board or plywood, also in bonding polyurethane foam to furniture frames.

The use of TCE expanded in the USA as it was exempted from the volatile organic compounds (VOC) regulations. Apart from the uses just mentioned it replaced ethyl acetate in adhesives for packaging; it was also applied in latex adhesives based on styrene–butadiene formulations. At present in Western Europe 75% of TCE production is used for metal cleaning, 10% in adhesives and 15% in aerosols, electronics and other applications. With concern about damage to the ozone layer there has been a change of policy and TCE is no longer favoured; it has an ODP of 0.1 (cf. p. 113) and consequently there is a move back towards the VOC listed above. This substitution may protect the ozone layer but release of these solvents enhances capture of NO (equation 5.11, p. 114) so further increasing tropospheric ozone levels.

Other protective measures include greater use of water-based latexes and emulsions. These are ineffective for the bonding of rubbers and some plastics; when sprayed they also produce fine mists which are carried around the workplace. Solid adhesives consisting of isocyanates and polyurethanes which cure when exposed to moisture are also available.

Solvent release may be avoided by the hot melting of plastics such as polyethylene and polyesters which bond to metals, plastics and paper on cooling. This technique is used in bookbinding, glazing and for the installation of car carpets and panels. Curing of acrylic, polyester and urethane resins by irradiation with UV, IR light or an electron beam is being used increasingly. These alternatives require expensive new equipment but are generally not more costly overall.

6.3.2 Coatings and inks

In 1986 almost half of coatings were solvent based. In 1989 this represented 1.7×10^6 tonnes but only 1.2% or 21 800 t was TCE. Its use was restricted to thinning for spraying inks on to wallpapers, and for labelling bags,

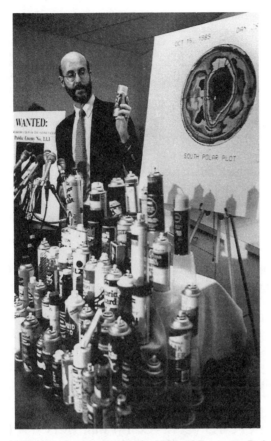

Figure 6.6 David Doniger, an attorney for the National Resources Defense Council, displays some of the 141 common household and office product aerosol cans that his organization says contain ozone-destroying chemicals. The products include hairsprays, spot removes and solvents. The cans contain 1,1,1-trichloroethane (methyl chloroform) (photo: Associated Press).

bottles and cartons. Here also the use of water-based materials and radiation curing controls solvent release.

6.3.3   Aerosol sprays

Figure 6.6 illustrates the much publicized concern about damage to the ozone layer arising from the use of CFCs in spray cans (p. 112). These formulations include three components: (i) the active ingredient such as insecticide, (ii) a solvent and (iii) propellent.

It has been estimated that in 1986 6.8 billion individual items were produced and many included TCE as solvent. At this time production in the USA was 18 600 tons and in Europe 12 400. In Japan production then

Table 6.12   Annual waste from cleaning car engine control boards

Waste type	Conventional soldering with CFC cleaning (kg/yr)	Controlled atmos. soldering (kg/yr)
Vol. organics	5400	900–1800
Formic acid	nil	45–90
Lead	4–7	0.5–5
CFCs	9000–23 000	nil
Work rate of moving belt	6 m/min	0.8–2.0 m/min

at 10 800 tons had fallen to 5000 by 1990. Alternative solvents are dimethoxyethane and low boiling alcohols, ketones and petrol fractions.

6.3.4   Metal cleaning

Trichlorethylene was phased out in 1960. At present CFCs are used for cleaning and drying a wide range of parts as also are perchlorethylene and methylene chloride. Table 6.12 shows how they can be eliminated from the waste from cleaning soldered car engine parts by the substitution of other organic solvents. A nitrogen atmosphere reduces the risk of fire but the throughput is also reduced. Other parts which require cleaning in solvent baths include adhesive spreaders, silk screen stencils, polymer formers, oil rig equipment, photocopiers and printing machines.

At present the SAAB company uses ethanol in place of CFC-113 for cleaning and Nissan (Japan) will replace CFC-11 (CCl_3F), CFC-12 (CCl_2F_2) and CFC-113 ($CCl_2F–CClF_2$) with HCFC-134 ($CF_3–CH_2F$) by the end of 1994. The nomenclature and ozone depletion potential (ODP) of these hydrofluoro-chlorocompounds is discussed on p. 113.

6.3.5   Dry cleaning of clothes

Organic solvents are used as, unlike water, they do not distort the fibres. The machines have long been totally enclosed on account of cost control. The solvents available are listed in Table 6.13. According to the provisions of the Montreal Protocol (p. 113) for protection of the ozone layer, CFC-113 (ODP 0.8) and TCE (ODP 0.1) will not be acceptable and should be phased out by 1997. This presents practical problems since dry cleaning equipment has a lifetime of about 15 years and is only suitable for the solvent originally specified. Even HCFC-225, the recommended substitute, will be unacceptable by the year 2020.

Carbon tetrachloride is still used for dry cleaning in Eastern Europe, the former USSR and in South East Asia. Otherwise it is produced by the high temperature chlorination of methane, propylene or carbon disulphide for

Table 6.13 Some solvents of use in dry cleaning

Solvent	Boiling point (°C)	Energy to boil 1 litre (kcal)
CFC-113	48	64
Perchlorethylene	121	116
Petrol based	150–210	—
1,1,1-TCE	74	90
HCFC-225 ($CF_3CF_2CHCl_2$)	53	64

use as a feedstock in the synthesis of CFC-11 (CCl_3F) and CFC-12 (CCl_2F_2). In 1988 the production in the USA of 283×10^3 tonnes was almost entirely used for this purpose. As these solvents are phased out it will only be used in a limited number of special syntheses, for example that of picloram (p. 238).

6.3.6 Solvent toxicology

The occurrence and toxicology of solvents in water has been extensively studied in the United Kingdom by the Water Research Centre based at Medmenham. Apart from protecting potable water supplies the analysis reveals those substances released by industry whether by evaporation to air or by direct discharge to rivers as waste.

Aliphatic hydrocarbons.   Some 60 compounds in this group have been detected in air and water consistent with their use in fuels. They do not in general present much risk to mammals but butane and *n*-hexane have attracted attention.

Butane accounts for 30% of all solvent abuse by glue sniffers and is obtained by them from lighter refills (Pottier *et al.*, 1992).

n-Hexane is one of the isomers present in motor fuel and in alkane mixtures used for paint thinning and for metal cleaning. It forms 1% of diesel emissions and 1.2% of the exhaust from petrol engines while losses from fuel tanks and carburettors can reach 10%. *n*-Hexane has a lifetime of about 6 hours in smog and has an odour threshold of 200–800 mg/m³.

On exposure for 10 minutes to levels of 2000 µg/g there was no acute response in man. The principal chronic effect in mammals and man from exposure is one of neurotoxicity.

Abnormalities in the nervous system have been noted in workers occupationally exposed and also in glue sniffers. *n*-Hexane is rapidly metabolized in mammals being excreted as the alcohol (**6**); its neurotoxicity is attributed to further oxidative metabolism to form 2,5-hexanedione (**7**).

Aromatic hydrocarbons. Attention is here focused on benzene. On inhalation acute effects include giddiness, headaches and nausea; chronic poisoning leads to depression of bone marrow function. There is evidence of its being a tumour promotor as it is linked to myeloid leukemia and lymphomas. The strength of economic factors can be judged from the fact that benzene has been banned as a laboratory solvent but it still forms up to 5% (vol.) of lead-free petrol. Over 15×10^6 tons are produced annually in England and it contributes to exhaust emissions (2.4%, vol. of total hydrocarbons). Near the M1 at Luton benzene reached a level of 155 ppb. The maximum allowable levels for an 8 hour day and 40 hour week range between 5 and 10 μg/g (16–32 mg/m^3).

Toluene is now preferred to benzene as a solvent. Death due to its inhalation has occurred as a result of solvent abuse and at high levels of occupational exposure (30 μg/g) it can induce mild abnormalities of the CNS. Levels in petrol and air are similar to those of benzene; its lifetime in smog is about 6 hours.

6.3.7 Organochlorine compounds

Vinyl chloride. This compound presents the most serious risk although it is still used in massive amounts for the production of PVC (Table 6.9). Its b.p. of -14°C makes for a problem of containment. Vinyl chloride (section 3.2.2) causes liver degeneration in mammals and is carcinogenic in man, inducing tumours of the liver and blood in those occupationally exposed; it is suspected as a cause of human mutations (Lewis, 1992). Accepted levels in the workplace are in the range 3–5 μg/g but may be made more stringent. Vinyl chloride is no longer available as a laboratory reagent.

Chloroform. This occurs in air at levels up to 5 μg/m^3 and has been detected in foodstuffs up to 30 mg/kg. It is common in water samples being produced by the haloform reaction during chlorination of water and sewage; in air it may be formed by the photochemical degradation of trichloroethylene.

Chloroform causes kidney damage and is cancer inducing in rats and mice. Studies showed that chlorination of tap water did not lead to enhanced chloroform levels in human blood plasma; the WHO has set a guideline of 30 μg/l for chloroform in drinking water.

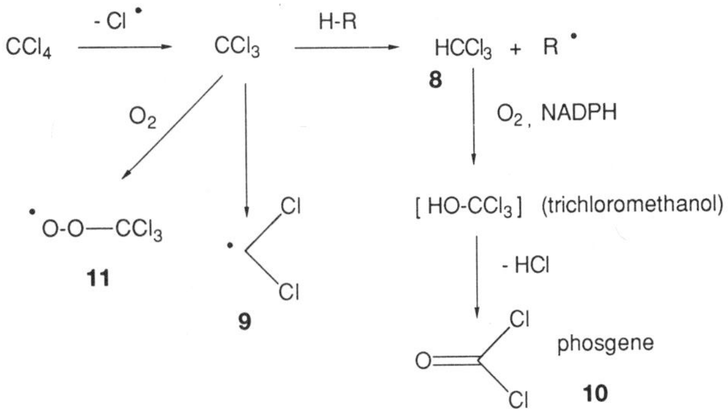

Figure 6.7 Metabolism of carbon tetrachloride.

Carbon tetrachloride.   As seen previously (p. 215), this is on the decline as an industrial chemical. It occurs in air and food; in drinking water it is at a level of 2–3 µg/l. Acute exposure in animals produces death by depression of the CNS or through liver damage. In the liver it is metabolized via the trichloromethyl radical (Figure 6.7) producing chloroform (**8**), dichlorocarbene (**9**), and phosgene (**10**), which is also a metabolite of chloroform. Capture of oxygen yields the long-lived peroxy radical (**11**) which is held responsible for liver damage.

Carbon tetrachloride causes liver cancer in animals and there is a possibility that this extends to humans.

1,1,1-Trichloroethane.   This is an important industrial solvent (p. 211). It occurs in air at about 1 µg/g and has been found in food and water samples. It does not appear to be metabolized in the manner of Figure 6.7, being largely exhaled from the lungs unchanged.

Deaths from acute occupational exposure have been recorded and solvent abusers are also at risk of heart failure. There is no evidence of long-term chronic effects but the accepted workplace level is 10 ppm.

6.3.8   Detergents

Detergents are a group of synthetic compounds within the general class of surfactants. They were first produced during the First World War in Germany but the period after the Second World War saw a dramatic escalation in their use. The principal products were based on cheap and readily available petrochemicals. The dramatic displacement of soaps by detergents is illustrated in Table 6.14.

A typical propylene tetramer (Figure 6.8, **8**) was combined with benzene in an acid-catalysed reaction. The position of the double bond and also the

Table 6.14 The displacement of soap by detergents in the USA (Davidsohn, 1987).

Year	Soap usage (10^3 t)	Detergent usage (10^3 t)
1940	1410	4.5
1950	1340	655
1960	583	1645
1972	587	4448
1982	545	5090

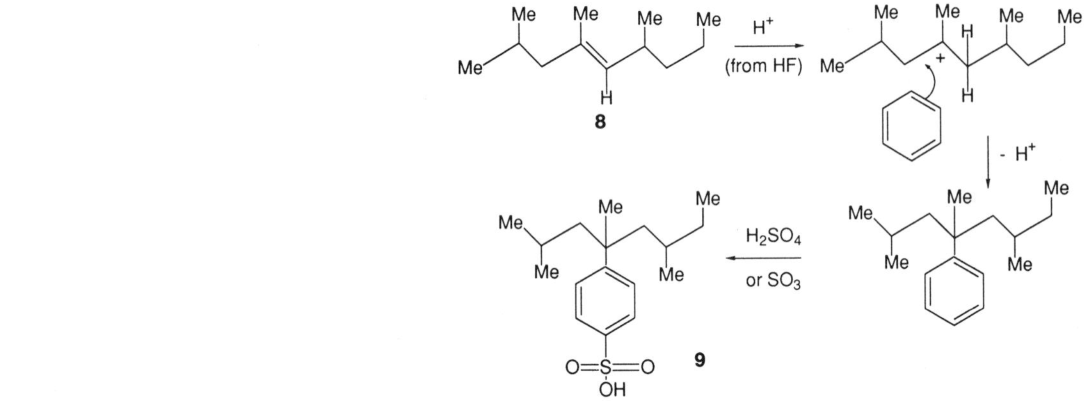

Figure 6.8 Synthesis of an alkylbenzene sulphonate.

number of polymerized alkene molecules varies because it is not economic to separate single substances, nor is a mixture of alkyl groups any less effective for detergency.

Types of detergent.   This sulphonate (**9**), like soaps of the stearate group (**10**) dissolves in water and becomes effective as its anion. These compounds are therefore known as anionic detergents. The sulphonates have the advantage that, unlike the stearates, they are not precipitated in hard water as calcium or magnesium salts. Both types depend for their action on removing organic material from soiled surfaces on the residence of the hydrophilic head in the aqueous and the lipophilic tail in the organic phase (Figure 6.9). This greatly reduces the surface tension allowing the non-aqueous materials to be wetted and washed out. Above a certain critical concentration the surfactant will support water-insolubles within aggregates or micelles, which present the heads to the water while the tails form an internal clump.

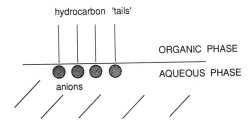

Figure 6.9 Phase interaction of an anionic surfactant.

Anionic detergents still occupy a dominant place in the market, but the use of detergents is widespread in industry and the home (KARSA, 1991). This necessitates the formulation of many specialist types including cationic (**11**) and non-ionic detergents (**12**).

CH_3-(CH_2)$_{16}$-C (=O) (O$^-$ Na$^+$)   A soap - sodium stearate

10

CH_3-(CH_2)$_{14}$-CH_2—N$^+$(Me)(Me)(Me) Cl$^-$ Hexadecyl trimethylammonium chloride used for home laundry and fabric conditioning

11

C_8H_{17}—⟨benzene ring⟩—O-CH_2-CH_2(-O-CH_2-CH_2)$_n$-O-CH_2-CH_2OH

12

when n=4 hexaethoxyethylene glycol monoether of *p*-octyl phenol

Builders.   In the original preparation of sulphonates, sulphuric acid in the residues was neutralized with alkali so forming sodium sulphate. This was retained so as to bulk up the product and make it easier to measure out. This practice continues and in addition other builders are added, including sodium carboxymethylcellulose (CMC), to prevent redeposition of dirt and phosphates to remove hardness from the water. Equation 6.13 shows how one of the phosphate formulations, tetrasodium pyrophosphate, sequesters magnesium ions so softening the water without precipitation:

$$2Na^+(Na_2P_2O_7)^{-2} + MgCl_2 \rightarrow 2NaCl + 2Na^+(MgP_2O_7)^{-2} \quad (6.13)$$

Pollution by detergents.   The upsurge in the use of anionic sulphonates based on propylene led to concern about persistence and foaming when they were released to water bodies. The traditional soaps are degradable by the effluent bacteria but the branched chains of the sulphonate are resistant. Consequently by the early 1960s industry began to formulate these detergents from degradeable straight chain polymers of ethylene, following the synthetic route shown above (Figure 6.8).

The production of sodium tripolyphosphate as an additive increased one hundredfold between 1947 and 1970 and levels in surface waters increased in proportion. Thus Lake Constance, situated in a populated area on the German–Swiss border, has no outflow and phosphate rose from a natural level of 0.2 mg (as P)/m^3 to 3 mg/m^3 by 1953 (Davidsohn, 1987): today this has risen by an order of magnitude. This type of accumulation of phosphate has supported the growth of algae and the associated unsightly 'blooms'.

To control algal growth alternative sequestering agents have been substituted for phosphate. One of these is the chelating agent nitriloacetic acid, which associates with calcium and magnesium as shown in structure **13** (cf. equation 6.13). Nitriloacetic acid (NTA) is, however, environmentally suspect since it also sequesters and therefore mobilizes heavy metals such as lead and cadmium from sediments. The most promising alternative to phosphate are zeolites which are polymeric sodium aluminium silicates. A simplified form is the structure **14**, where the trivalent aluminium atoms in sharing with the tetravalent silicon acquire a surplus negative charge; as a result the zeolite associates wiith calcium ions in order to achieve electrical neutrality.

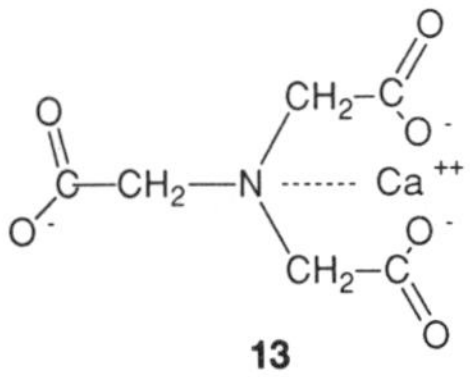

13

association of sodium nitriloacetic
acid with calcium ion

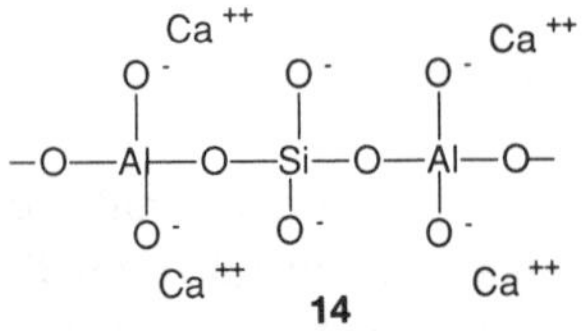

14

simplified structure of a zeolite

6.3.9 Indoor pollution (see also p. 23)

It is not always appreciated that pollution levels in the home and workplace may exceed those in the urban environment outdoors (see National Academy, 1981). This is a matter for concern as individuals spend 90% of their time indoors.

Volatile organic compounds (VOCs).   Indoor air contains a complex mixture derived in large part from the use of propellants (p. 213) to dispense cleaning compounds, deodorizers and foodstuffs. To this source must be added gases from cooking and heating fuels, refrigerants, adhesives, dry-cleaning solvents, textiles, floor coverings, dyestuffs and pesticides. As mentioned previously (p. 215) 1,1,1-trichloroethane is one component and tri- and tetra-chloroethylene contribute. Alkanes and alkylated aromatic hydrocarbons also occur together with their oxidation products (cf. equation 6.9).

Hydrocarbons other than methane may exist at levels up to 8.0 μg/g as compared with a range of up to 3.5 μg/g outdoors. High levels in the range of 10–50 μg/g are common in new offices and school buildings; these originate from new furnishings and fittings.

Formaldehyde is the most important of the indoor carbonyl pollutants. It is released by the outgassing of urea–formaldehyde foam insulation from particle board, cigarettes and other indoor combustion especially gas cookers. Typical indoor levels range between 0.05 and 0.3 μg/g

Nitrogen oxides (NO_x).   These arise from ineffectively vented gas cookers and heating systems. In these circumstances levels frequently exceed the air-quality standards for outdoors and may fall in the range 20–300 ng/g.

Carbon monoxide (CO).   This is prevalent in underground garages and is also produced indoors by gas heaters and cookers. Levels indoors are increased by cigarette smoking and are usually higher there than outside, in the range 0.5–5 μg/g with a peak at 25 μg/g. Smoking also introduces acrolein (CH_2=CH–CHO), benzo-[a]-pyrene and trace metals.

Other gaseous oxides.   Carbon dioxide levels may reach 2000 μg/g, five times that outdoors, but sulphur dioxide is usually lower indoors at < 20 ng/g. Ozone is also usually at a low indoor level in the range 0–20 ng/g.

Analysis of the complex mixture of indoor pollutants requires refined techniques including capilliary GLC (p. 70, Bayer and Black, 1987).

6.4 Organohalides: pesticides, PCBs and dioxins

6.4.1   Historical

Chlorine was first isolated by the Swedish chemist Carl Scheele in 1774 and its commercial use as a bleaching agent dates from 1789. In 1800 Cruickshank showed that it could be obtained by the electrolysis of brine and following the development of the mercury and diaphragm cells during 1883–93, this method is now employed for large-scale production. The mercury cell reactions are:

$$Na^+ + Cl^- \rightarrow Na/mercury + Cl\cdot$$

whence sodium discharged at the cathode reacts with water to produce sodium hydroxide and hydrogen. Chlorine gas is collected from the anodic discharge.

Mercury is surprisingly soluble in water (10^{-1} mg/l at 50°C) and as a result losses amount to between 100 and 200 g/ton of chlorine produced. Owing to the toxicity of mercury (p. 157) and more stringent controls on its release this has necessitated the use of diaphragm cells in industry. In this modification the chlorine stream at the anode is separated from hydrogen and sodium hydroxide by a layer of asbestos (p. 189) which is vacuum-deposited on the cathode. The brine now percolates this diaphragm under close control to enter the cathode compartment.

In addition to its use in bleaching, large quantities of chlorine were required for use as an insecticide. One example of the importance of this application is the elimination of cholera in developed countries as a direct result of chlorination of water. In the UK the first recorded death from this loathsome disease was that of William Sproat in 1832 and about 32 000 deaths occurred here then in a two-year period. Cholera was pandemic during 1862–1872 and came under control following identification of the bacillus by Koch in 1884. The last recorded outbreak in western Europe was that in 1892 in Hamburg when 10 000 died, but 1000 deaths in Peru were attributed to cholera in 1991.

6.4.2 Organochlorine production

Despite its use in water treatment, worldwide chlorine production was relatively modest at about 2 million tons in 1940 but this increased rapidly with the demand for organochlorine compounds for use as pesticides. Figure 6.10 shows the exponential growth in the synthesis of chemicals from petroleum and the related demand for chlorine in the USA. By 1971 petrochemicals amounted to over 90% of the total in the USA, Europe and Japan; chlorine consumption has now levelled out partly as a result of restrictions on polychlorinated pesticides (pp. 229–30).

The first compound produced commercially was carbon tetrachloride in 1907 followed by others such as trichloroethane in the 1920s. Annual production worldwide now ranges from over 10^9 kilograms of industrial solvents such as dichloroethane down to a few million kilograms of more specialized substances, for example hexachlorobenzene.

6.4.3 DDT (dichlorodiphenyl trichloroethane)

Apart from the ready availability of chlorine and petroleum products in the years following the Second World War, the successful use of DDT (Figure

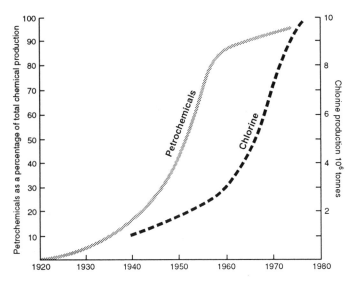

Figure 6.10 Comparative growth of petrochemical and chlorine production in the USA 1920–1975.

6.11) in controlling disease during its final stage encouraged the research and development of organochlorines as a whole. DDT was first described by Othmar Zeidler in 1874, although over sixty years elapsed before it was formulated as an insecticide by the Geigy company following the work of Muller (1940).

The preparation of DDT depends on the condensation of chloral with chlorobenzene in sulphuric acid (Figure 6.11). This is not a unique pathway and significant amounts of the isomer with one chlorine in an *ortho*-position are produced; also chloral produced by the chlorination of acetaldehyde is contaminated by dichloroacetaldehyde and gives rise to DDD. Commercial DDT has been shown to be a mixture of 14 substances – DTT itself constitutes 65–80%; *o, p*-DDT 15–20% and DDD may be up to 4%.

6.4.4   Lindane, hexachlorocyclohexane (Figure 6.12)

This substance is obtained by the addition of three molecules of chlorine to benzene activated by ultraviolet irradiation. In theory there are eight possible geometrical isomers in which the chlorine atoms occupy different relative positions about the cyclohexane ring. Structure (**15**) is drawn in a way which shows the characteristic 'chair' form with six axial (a) bonds parallel to the main axis and six equatorial (e) bonds disposed around the belt of the molecule. The active form, the *gamma*-isomer, has three consecutive axial and equatorial substituents and forms about 15% of the

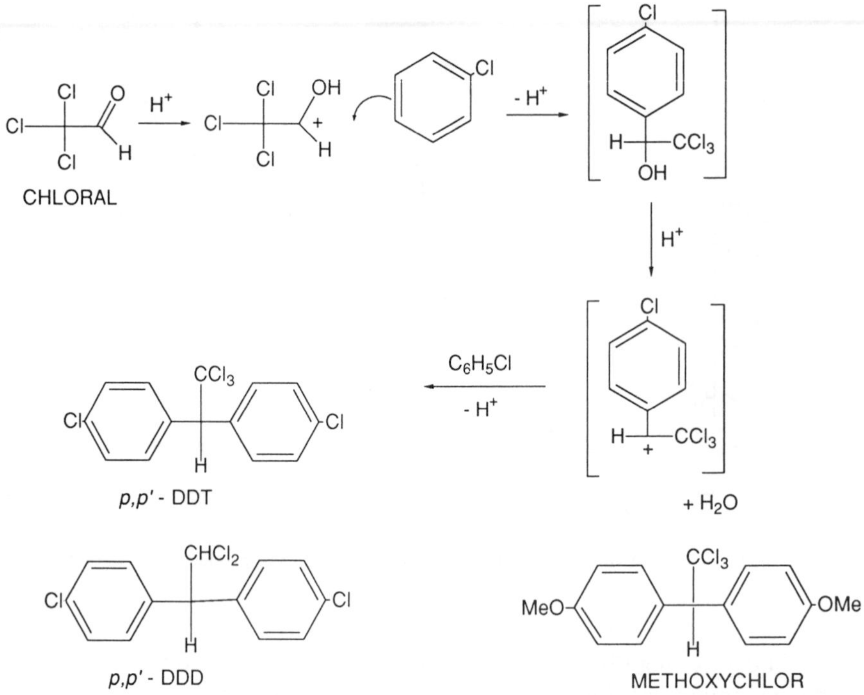

Figure 6.11 The preparation of DDT.

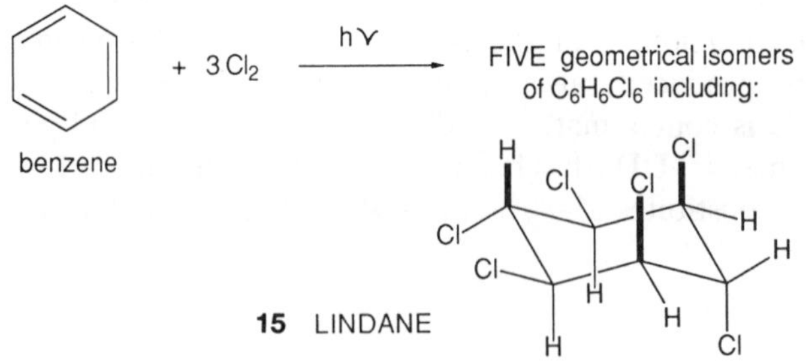

Figure 6.12 The synthesis of Lindane.

mixture of reaction products which includes five of the possible isomers. It was first described by Van de Linden in 1912 and introduced commercially by ICI in 1942.

Lindane or gammexane has similar properties to DDT and is widely used in Third World countries as the crude product is cheap to produce although some of the components have a musty taste. For control of pests in food crops such as the potato it is necessary to purify lindane, which is tasteless, by recrystallization. It is superior to DDT in controlling soil pests.

6.4.5   Some other chlorinated pesticides

The structures of some other principal chlorinated pesticides are shown in Figure 6.13. A number of these compounds were obtained by the Diels–Alder addition reaction of hexachloropentadiene (**16**). Combination with bicycloheptadiene (**17**) leads to Aldrin (**18**), which on epoxidation affords Dieldrin (**19**). A similar relationship is found between Heptachlor (**20**) and its more active epoxide (**21**).

Epoxidation is an important *in vivo* process and results in the hydroxyla-tion and consequent solubilization of organic pollutants (p. 242). While this change is the key to degradation of toxic compounds there are occasions when the cure is worse than the disease. Thus Aldrin is

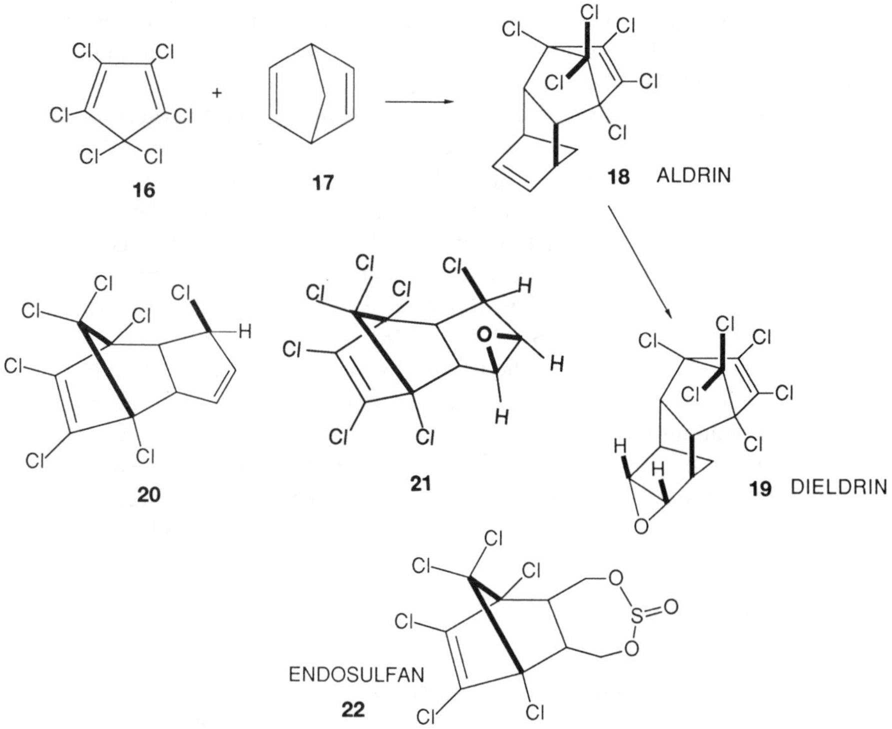

Figure 6.13   Synthesis of the Aldrin group of pesticides.

Table 6.15   Toxicity to fish (96-hour static LC_{50} ($\mu g/g$))

	Aldrin	Dieldrin
Striped mullet	100	23
Bluehead	12	6
Rainbow trout	36	19

converted into Dieldrin *in vivo* and the latter is in general more toxic to fish (Table 6.15). The concentration of residual chemicals in food chains is a further cause for concern and has led to restrictions on the use of these compounds (p. 229).

The cyclic sulphite Endosulfan (**22**) is still in use, even in the USA, since the high level of oxygenation ensures that it is less persistent than other members of this group.

6.4.6   Organochlorine herbicides (Cremlyn, 1991)

Early non-selective chemical control agents included oil waste and creosote. A degree of selectivity was achieved by spraying cereal crops with copper sulphate solution or dilute sulphuric acid, when the large broad-leafed weeds were more susceptible.

1933 saw the first use of a truly selective agent, dinitro-*o*-cresol (**23**, Figure 6.14) and although limited by its failure to control perennial weeds it is still useful as a winter wash for fruit trees and is not persistent. It is, however, very poisonous to mammals with an LD_{50} of 30 mg/kg in rats.

The discovery by Kogl in 1934 that 3-indolylacetic acid (**24**) was a plant growth hormone led to the development of more active and more stable compounds for crop protection during the Second World War. Notable among these were 2,4-D (**25**), 2,4,5-T (**26**) and MCPA (**27**).

These substances were applied typically at a level of 1 kg/ha to broad-leafed weeds in grassland. They act as mimics for the natural growth hormone leading to over-production of RNA and death of weeds, which cannot then obtain sufficient nutrients from the roots to sustain the abnormal growth. Demand is decreasing because of resistance, the emergence of alternatives and the presence of toxic by-products in the formulations.

An aliphatic substance, Dalapon (**28**) is useful for the control of couch grass. This agent is not persistent as it is readily hydrolysed to pyruvic acid (Figure 6.14).

Despite their cheapness and early dramatic successes undesirable pollution has arisen from the use of the organochlorine pesticides. This has led to restrictions on their use and a search for more expensive alternatives (p. 244).

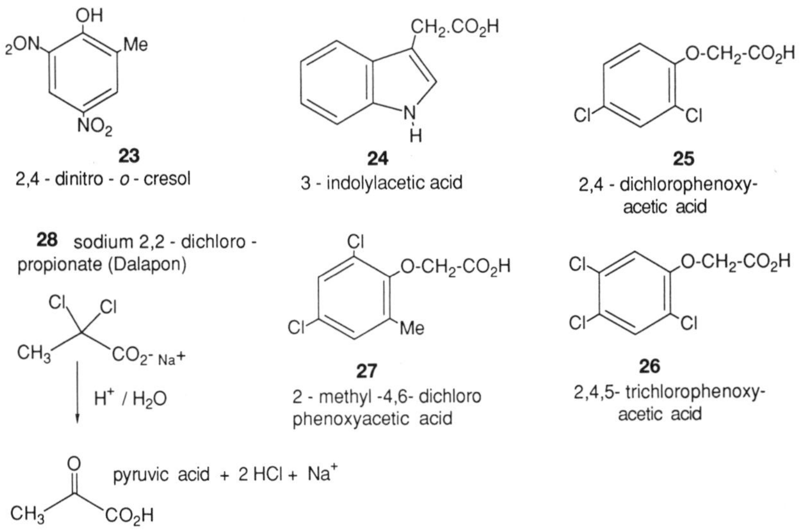

Figure 6.14 Some important herbicides.

6.4.7 Toxic effects of insecticides

It can be appreciated that to assay the toxicity of pure individuals in mixture arising from commercial synthesis may be difficult: commercial samples of DDT include 14 substances (p. 223). The PCBs provide an extreme example of this; they are formed as a mixture arising from the mechanism of their synthesis rather than from impurities in starting materials (p. 233).

DDT is not very toxic to man – a human test group ingested 35 mg daily over an extended period without ill effects – however, a typical insect toxicity would be LC_{50} of 5 µg/l for a 48 hour exposure. It is extremely toxic to cold blooded creatures owing to their inability to limit the opening of sodium ion nerve channels (cf. p. 249) by the wedge-shaped DDT molecule; as a result nerve cells are continuously activated and death follows.

Penetration of the insect cuticle is favoured by high levels of molecular chlorine, this enhances the effectiveness of chlorine-rich molecules such as DDT ($C_{14}H_9Cl_5$). Unfortunately, the survivors of successive generations of pests breed ever more resistant strains and by 1950 several species of house fly and of cabbage rootfly had become resistant to DDT. This is not likely to have arisen from induced mutations but rather from the survival of a small resistant proportion of the insect population followed by the relatively rapid breeding of these more resistant survivors (Hutson and Roberts, 1987). This process of recovery is often accelerated by the action of high doses and the reduction of predator populations, which recover their numbers more slowly. By 1980 over 400 species of insects and the very rapidly reproducing mites had bcome resistant to DDT. Related compounds like methoxychlor acquire resistance by a crossover mechanism but this does not extend to compounds in a different cross-resistant group (Figure 6.11) such as Aldrin (**18**) and Lindane (**15**). To overcome resistance, application rates have to rise and the search for effective alternative treatments is ongoing. It is also found that the introduction of electro-negative chlorosubstituents makes organic compounds more difficult to oxidize (p. 242) and so extends their lifetime, which for DDT may extend to >10 years.

Largely as a result of unwise excessive applications the use of DDT and its analogues was called into question in developed countries. This followed the realization that fish and raptorial birds at the head of food chains were being poisoned owing to biomagnification along the chain.

Figure 6.15 shows relationships in the complex ecosystem of Lake Kariba, Zimbabwe, with clear biomagnification of sedimentary DDT levels to those in filter feeding *Corbicula africana*. The relation between kapenta and algae is not clear, possibly because the short-lived plankton reflect DDT leveis over a short time period while the levels in kapenta have been accumulated over a much longer period. Biomagnification is clearly shown from kapenta to the tigerfish and the cormorant. Other evidence of biomagnification is seen between fish and the crocodile at the head of the chain.

Lack of understanding of food chain relationships has led to numerous pollution incidents which became the subject of public concern.

Clear Lake, California (Moriarty, 1988). This is a body of water about 100 square miles in area, 70 miles north of San Francisco and much used for recreation. Following complaints about clouds of gnats the lake was first treated with DDD in 1949 and the nuisance was controlled; following recovery in the numbers a successful reapplication was made in 1954. However, by 1957 the gnats and numerous other insect species had developed resistance to DDD. Towards the end of 1954 many dead western grebes were found, leading to a public outcry which ultimately ended the treatment. These birds were fish-eating divers and their death was attributed to bio-accumulation through plankton → plankton-consuming fish → carnivorous fish → grebes. The DDD was originally dispersed in the lake water at a concentration of only 20 ppb, but the birds had accumulated DDD in their fat at levels up to 1600 ppm – a bio-accumulation factor of 80 000.

Death of predatory birds. A small fall in the peregrine population around 1940 (Figure 6.16, Cremlyn, 1991) was the result of deliberate

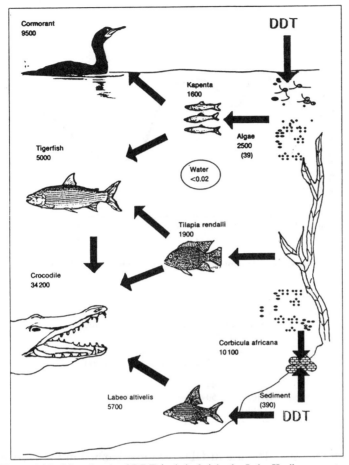

Figure 6.15 Mean levels of DDT (ng/g in fat) in the Lake Kariba ecosystem (Berg, 1992).

shooting in wartime to protect carrier pigeons. Post-war recovery was followed during 1950–1960 by a steep and clearly abnormal decline, which was attributed to the application of organochlorines, especially Aldrin and Dieldrin, to protect cereal seeds from the wheat bulb fly. Other toxic effects were seen among pigeons, other predatory birds and foxes at the head of the food chain.

DDT and other organochlorines are responsible for the thinning of eggshells and the consequent reduction in hatching rates.

6.4.8 Control of pesticides

With increasing awareness of the toxic effects of chlorinated agrichemicals limitations have been set on their use and in some instances a complete ban

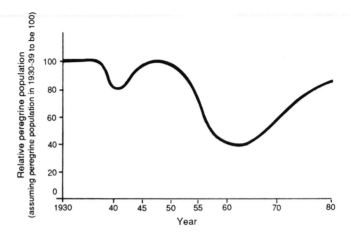

Figure 6.16   UK populations of the peregrine falcon (1930 = 100).

Table 6.16   Persistence of some pesticides in soil (Goring, 1972)

	% Decay	Time taken (years)
DDT	44	8
BHC	50	1
Dieldrin	49–53	3

has been imposed. The use of DDT was first banned in the USA in 1973 and a ban on Aldrin and Dieldrin followed in 1983. In the UK DDT was banned in 1984 and the use of Aldrin and Dieldrin for seed dressing was stopped in 1986. However it must be emphasized that even following prohibition these substances will persist in the environment for a period dependent on temperature, rainfall and soil type; some median half-life values are given in Table 6.16.

It should also be noted that insects amount to 70% of all animals and of one million insect species about 1% are significant pests, so alternatives must be found. In developed countries the policy has been to reduce dependence on organochlorines and to introduce environmentally acceptable but more expensive substitutes such as pyrethroids (p. 244) and organophosphorus compounds (p. 248).

In Third World countries a ban on effective control chemicals is not enforced as rigorously as in the developed countries since the magnitude of the problem posed by pests overshadows possible toxic effects. As an illustration it should be noted that 100 million clinical cases of malaria are reported annually and of these about one million are fatal. A problem compounded by the fact that fifty of the sixty species of malarial mosquitoes are resistant to DDT and Dieldrin. Another instance is the distribution of the tsetse fly over 12 million square kilometres of Africa with the associated risk of trypanosomal disease in cattle and humans.

The problem of contamination of herbicides by dioxins is closely related to that arising from the oxidation of PCBs and is discussed in the following section.

6.4.9 Vinyl chloride and polyvinyl chloride (Kirk-Othmer, 1983)

In this section, and that following, pollution arising from accidental leakage of organochlorines produced for applications other than control of pests is considered.

Some 85% of synthetic chemicals are used for the manufacture of polymers. In 1985 about 7 million tonnes of vinyl chloride were produced worldwide by a route dependent on the addition of chlorine to ethylene and the subsequent thermal elimination of hydrogen chloride. To avoid wastage of hydrogen chloride acetylene has been added to the system to trap it and so provide a balanced process:

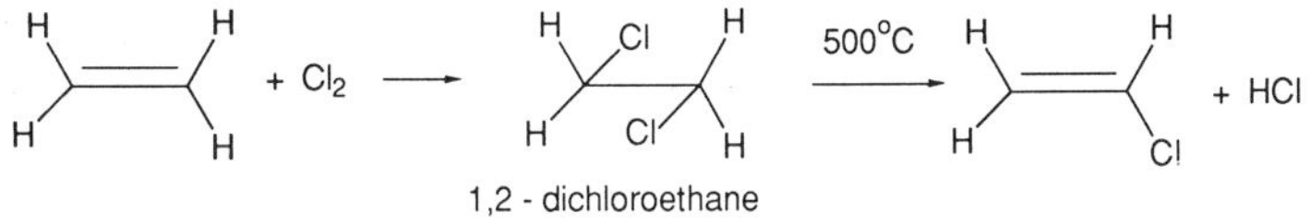

1,2 - dichloroethane

$$C_2H_4 + C_2H_2 + Cl_2 \rightarrow 2\ CH_2=CH.Cl \qquad (6.15)$$

More recently re-incorporation of hydrogen chloride is ensured by oxidative addition, followed by thermal cracking:

$$C_2H_4 + 2HCl + [O] \rightarrow \underset{\substack{| \\ Cl}}{CH_2} - \underset{\substack{| \\ Cl}}{CH_2} + H_2O \qquad (6.16)$$

The production of polyvinylchloride (PVC) began in 1940 through the induction of polymerization of the monomer by free radicals. The reaction is conducted on an aqueous suspension in glass-lined steel kettles and a typical catalyst is benzoyl peroxide (29) in which the weak central bond is cleaved at 50°C to generate two initiators:

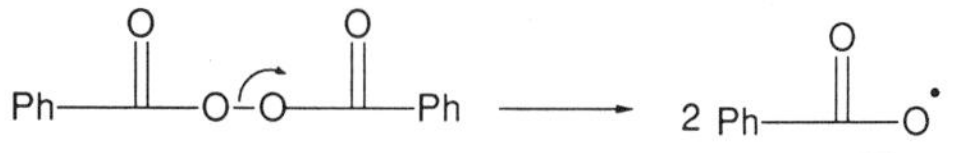

29

(6.17)

$$Ph-\overset{O}{\overset{\|}{C}}-O\overset{\frown}{} CH_2{=}\overset{\frown}{CH.Cl} \longrightarrow \left[Ph-\overset{O}{\overset{\|}{C}}-O-CH_2-\overset{H}{\underset{CH_2=CH.Cl}{\overset{\cdot}{C}}}-Cl \right] \qquad (6.18)$$

and growth continues in this way to give a degree of association of about 1000 monomer units and PVC granules with a molecular weight of about 63 000.

It is often not appreciated, especially by the media, that very few chemicals have been proved to cause cancer in man (p. 48). However, vinyl chloride is a carcinogen and by 1970 it was shown that industrial workers when exposed to it were prone to cancer. The causative agent is an oxidation product, chloroethylene oxide, which provides another example of toxicity emerging on modification in the environment (cf. Dieldrin, p. 225). 10% of ethylene is converted into PVC to meet a major demand for which no suitable substitute exists and hence there has been no ban on the monomer, although control measures have been adopted and it is no longer available as a laboratory chemical.

6.4.10   Polychlorobiphenyls (PCBs)

These compounds were first produced industrially in 1929 and were early on incorporated in printing inks and paints; subsequently their use was extended to the softening of plastics and as insulators in transformers. In the USA production peaked at 7×10^4 tons in 1970 and it is estimated that 75×10^4 tons have been manufactured overall. Their thermal stability made them attractive as insulating transformer fluids and biphenyl (30), the essential precursor, was readily available, as was chlorine. The chlorination (Figure 6.17) is catalysed by iron (III) chloride but exhibits the characteristics of a free radical reaction in that a diversity of biphenyls at different chlorination levels appear shortly after reaction begins. As a result it is not economic, or even feasible, to separate individuals and the mixed products are marketed under trade names such as Aroclor, Phenoclor and Clophen. The Monsanto nomenclature typically designates Aroclor 1242 as the fraction containing 42% of chlorine, b.p. 325–366°C; Aroclor 1268 contains 68% of chlorine, b.p. 435–450°C.

Of the three possible initial products the 4-chloroisomer (31) predominates with a significant amount of the 2-chloroisomer (32) and a little of the 3-chlorobiphenyl (33). Each of these compounds subsequently and rapidly captures chlorine in either ring A or ring B, so that substances such as 2,2',3,4'-tetrachlorobiphenyl (34) and others of higher chlorine number occur in Aroclors. There are 42 possible tetrachloroisomers and also 42 hexachloroisomers, because in the latter the four unsubstituted positions permute in the same way as do the four chlorinated ones. In total there are 209 PCBs ranging from chlorine number one to the fully substituted

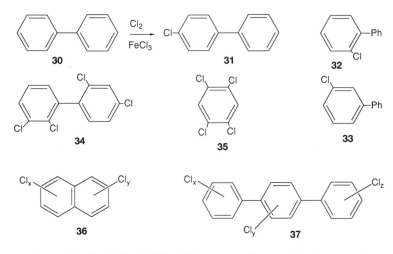

Figure 6.17 Synthesis of PCBs with structures of some related compounds.

Table 6.17 Occurrence of PCBs

	ng/m^3		mg/kg
Air	0.1–20	Plankton	0.01–20
Water	0.1–3000	Invertebrates	0.01–10
Sediments	1.0–1000	Fish	0.01–25
		Birds' eggs	0.1–500
		Man	0.1–10

decachlorobiphenyl. Hence the separation of all the pure components and evaluation of all the individual toxicities was a nearly impossible task. However, PCBs were identified as pollutants in 1966 with similar behaviour on gas chromatograms to that of DDT and like it they are now ubiquitous in the environment. Table 6.17 shows some typical occurrences.

It should also be pointed out that other chlorinated aromatic compounds constitute an environmental risk. These include chlorobenzenes (e.g. **35**), the chloronaphthalenes (e.g. **36**) and polychloroterphenyls (e.g. **37**). The last named are found in the environment at lower levels than PCBs but levels in human fat are comparable.

At normal levels of exposure PCBs are not very toxic to man, although in the Yusho incident their inclusion in rice bran cooking oil led to serious consequences. Residents in South Japan became affected by a peculiar skin disease which first broke out in March 1968; it became notorious as 'Yusho disease' and another outbreak occurred in Taiwan in March 1979 when about 2000 people were affected.

In 1973 a community of several thousand in Michigan was also seriously affected by the inclusion of polybromobiphenyl in cattle feed. This product

Table 6.18   Pollution indicators and incidents involving PCBs

1970–1976	Reduction in herring gull populations on the Great Lakes Erie and Ontario, which are most at risk from anthropogenic sources.
1950 onward	Decline in the populations of the European otter.
1950–1970	Decline in seal populations in the Baltic and Dutch Wadden sea (Harrison, 1992). Levels of penta- and hexachlorobiphenyls in the range of 2–15 ppb in Jordanian soils; these are similar to those found in industrialized countries (Alawi, 1991).
1987–1988	Over 700 dolphin carcasses were found on the Atlantic coast of the USA. The deaths were attributed to brevitoxin, a neurotoxin produced by the Florida red tide organism (*Ptyodiscus brevis*) but stress from the ingestion of organochlorines was a likely factor (Kuel *et al.*, 1991). Blubber levels in ppm:

	max.	mean
PCB	195	40
DDE	80	8
Dieldrin	3.3	1

1991	Rural air samples from Ulm, Germany, typically contain 170 pg/m³ of PCBs similar to those found elsewhere in Europe and over the Great Lakes (Ballschmitell, 1991). Air levels in Reunion, South Indian Ocean, were as high as 125 pg/m³. It is evident that the principal input to the northern hemisphere disperses slowly into the southern hemisphere (Ballschmitell, 1991).
1992	Milk from five species of seals was sampled from feeding regions in the Arctic, Antarctic, California and Australia. All contained organochlorines and PCBs, which were principally a common pair of hexachloroisomers and a sole heptachloroisomer. Typical levels (ppm) in the milk were: Australia 44 Arctic 232 California 370 Antarctic 10 (Bacon *et al.*, 1992)

consists mainly of penta- and hexabromobiphenyls, is used as a fire retardant and is marketed as 'Firemaster'.

Table 6.18 summarizes some observations and incidents which establish that pollution by PCBs is serious and widespread. Once the distribution of PCBs and the similarity of toxic effects to those of DDT were established, their use declined and was banned for all but closed systems by the Dow company in 1971. Production worldwide had practically ceased by 1977 but the problem of destruction of the residues remained.

The incident at Binghampton.   There have been numerous examples of pollution arising from transformer fires but that in the Binghampton State office building on 5th February, 1981 rates as the most serious in terms of damage to the building, state records and human exposure (Schechter *et al.*, 1985).

The transformer contained Aroclor 1254 and tetrachlorobenzene (35), the combustion products entered the ventilation system and hence were distributed throughout. Some hundreds of workers engaged in the clean-up were exposed to risk before the full extent of contamination was discovered and two years after the fire the building was still unusable. Table 6.19 lists the principal findings.

Table 6.19 Pollution of the Binghampton Office Building

Initial levels* (ppb)		Human adipose levels PCDD + PCDF (ppt)	
10% of soot equiv. to	100×10^6	Highest recorded	8400
of which PCDF	2.1×10^6	Arithmetic mean	2400
PCDD	$1–2 \times 10^4$	Mean from four people not exposed	1150
Early PCB in air	$80\ \mu g/m^3$		

* General standard of detection 1 ppt
PCDD = polychlorodibenzodioxins; PCDF = polychlorodibenzofurans

This incident graphically illustrates the toxic mix which results when organochlorines are burnt below the oxygen levels and temperatures needed for safe incineration (p. 242). The formation of a PCDF during PCB combustion probably involves the hydroxyl radical and the mechanism of the reaction is shown in Figure 6.18. 3,3′,4,4′-tetrachlorobiphenyl (37) is taken as an example and the formation of the free radical (39) from it follows a similar course to that shown in the conversion of methane to PAN (p. 203). The next step gives the phenoxy radical (40) and is followed by attack within the same molecule with closure of the furan ring and loss of a hydrogen atom. The product, 2,3,7,8-tetrachlorodibenzofuran (41), is the most toxic of this group and compares with 2,3,7,8-tetrachlorodibenzo-dioxin in the dioxin group. The TCDFs were implicated in the rice oil incidents and further examples of their toxicity are discussed in the section covering pollution by herbicides (p. 238).

It should also be noted that PCB fires generate chlorobiphenylenes such as (42) by dimerization of the radical intermediate (39).

6.4.11 Toxic substances in herbicides (Kamrin and Rodger, 1985)

An alternative title to this section would be 'the dioxin problem'. In the late 1960s chemists became aware of the presence of the dioxin analogues of the PCDFs (e.g. 41); of these the most dangerous is the symmetrical 2,3,7,8-tetrachlorodibenzdioxin (45). It was realized that this compound was formed as a by-product whenever the temperature of a reaction mixture containing 2,4,5-trichlorophenol exceeded 220°C. This explains

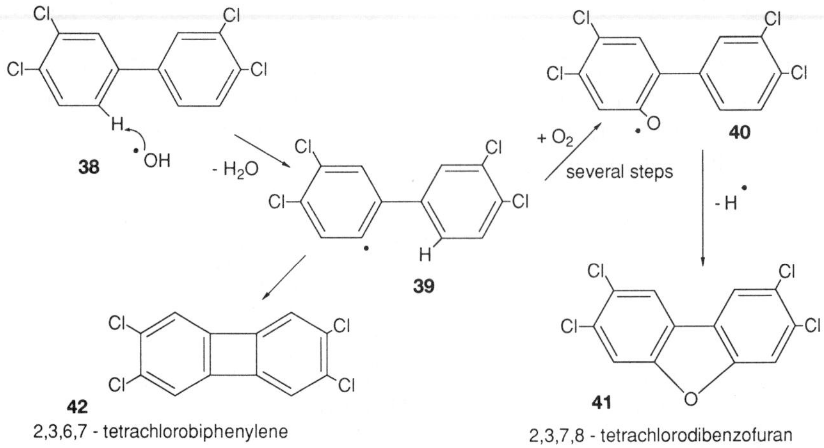

Figure 6.18   The formation of a PCDF from a PCB.

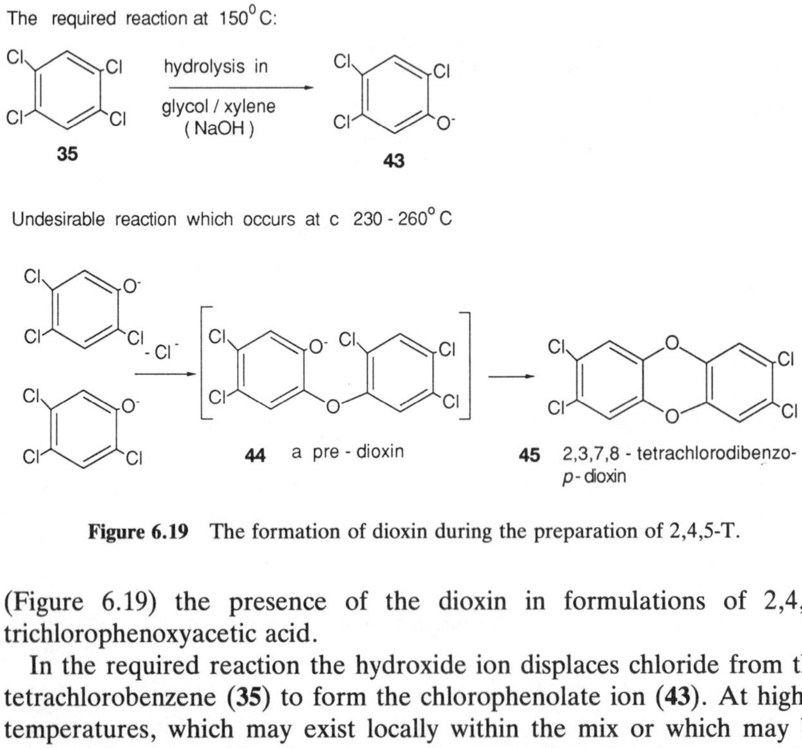

The required reaction at 150°C:

Undesirable reaction which occurs at c $230 - 260^{\circ}$ C

Figure 6.19   The formation of dioxin during the preparation of 2,4,5-T.

(Figure 6.19) the presence of the dioxin in formulations of 2,4,5-trichlorophenoxyacetic acid.

In the required reaction the hydroxide ion displaces chloride from the tetrachlorobenzene (**35**) to form the chlorophenolate ion (**43**). At higher temperatures, which may exist locally within the mix or which may be reached if controls fail, then the attack of one phenolate ion on another begins to compete. The first product is the diphenyl ether (**44**) but this rapidly cyclizes in a manner formally analogous to the last step in PCDF formation.

Table 6.20   Some dioxin pollution incidents

1949	A kettle used for trichlorophenol production at the Monsanto plant in West Virginia blew and exposed workers were affected by chloracne, a severe and persistent form of acne known to be typical of contact with 2,3,7,8-TCDD. (Kamrin and Rodgers, 1985)
1976	On 10th July overheating of a reactor at the ICMESA plant at Séveso near Milan led to the release of a cloud of solvent containing trichlorophenol and TCDD. About 1800 ha were contaminated and birds, animals and vegetation were seriously affected (Pocchiari *et al.*, 1983).
1977	Massive quantities of 'Agent Orange', an equal mixture of 2,4-D and 2,4,5-T, were sprayed from aircraft in the 1960s to defoiliate supply trails used by the Vietnamese during the war with the USA. In the summer of 1977 8 million litres of unused herbicide were destroyed by incineration at sea. It was estimated to contain 23 kg of 2,3,7,8-TCDD (Young, 1988).
1982	Deaths of horses and birds at a horse arena in eastern Missouri were attributed to the spraying of dioxin-containing waste oil to keep down dust. Humans were affected and the source was identified as a chemical plant in Verona, Missouri that had produced trichlorophenol for synthesis of the anti-bacterial agent hexachlorophene during 1970–1971. Fish in nearby streams were also found to contain 2,3,7,8-TCDD (Kleopfer and Kirchmer, 1985). All the streets in the small town of Times Beach, MO, were sprayed with the waste oil and it had to be closed down at a cost of $33 million. In 1991 the total cost of clean-up in Missouri was $80 million.
1981	Local authorities in the vicinity of Chesterfield, Derbyshire were concerned at the possible pollution of water supplies arising from historic dumping of chlorophenolic wastes by the Coke and Coalite Company.
1982	In a similar incident a former dumpsite at Love Canal, Niagara, USA was found to be polluted. The canal was excavated in 1894 with a view to linking the upper and lower Niagara rivers but was never completed and it was used to dispose of chlorophenolic wastes produced by the Hooker Corporation. The site was subsequently developed as a residential area and later when the risks were realized storm sewer sediments were found to contain up to 600 ppb of 2,3,7,8-TCDD (Kamrin and Rodgers, 1985; cf. p. 274).

There is no obvious route from PCBs to dioxins but they may still be formed in transformer fires as 1,2,4,5-tetrachlorobenzene is often added to keep the mix fluid. Clearly replacement of chlorine by oxygen during combustion is all that is required to generate dioxins in the manner shown in Figure 6.19.

The formation of the dioxin (**45**) occurred widely and as a result it became incorporated in industrial wastes that were frequently handled without knowledge of the risk. The formation of highly toxic dibenzodioxins and dibenzofurans during incineration is discussed on p. 243. The incident at Séveso and the Agent Orange issue (Table 6.20) are of first importance and will now be treated as case histories.

Agent Orange.   This was the major material used for defoliation by the USAF during the Vietnamese war (Young, 1988); others used early on

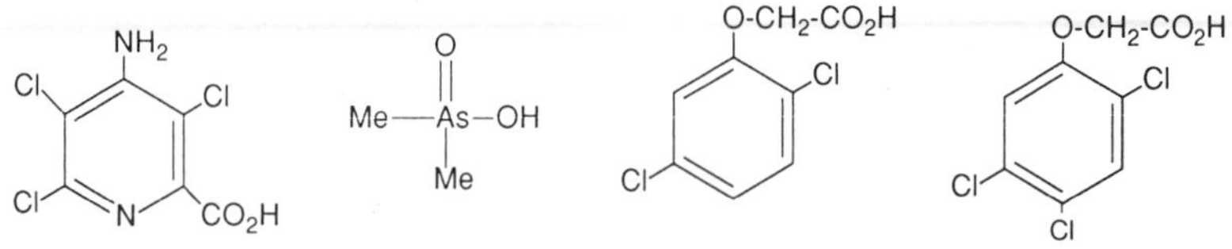

Figure 6.20 Herbicides used in Vietnam – all operations.

were codenamed Agents White and Purple. The mixture of herbicides in Agent Orange contained equal amounts of 2,4-D and 2,4,5-T (Figure 6.14) and variable amounts of dioxins, depending on the source and the degree of control during preparation. The levels of 2,3,7,8-TCDD ranged between 2 to 50 μg/g. This was not appreciated when the programme began in January 1962 but opposition in the USA grew until following an investigation by the National Institute of Health in 1969 the last mission was flown in May 1970.

In fact the preparation of 2,4-D does not present the same risk as that from 2,4,5-T as the side reaction of 2,4-dichlorophenol can only lead to dioxins of chlorine number less than four; this group is not dangerously toxic. It is worthy of note that the risk is now defined as a toxic equivalent factor (TEF) expressed as a fraction of 2,3,7,8-TCDD taken as unity. On this scale other dangerous individual compounds are 1,2,3,7,8-pentachlorodibenzdioxin (TEF = 0.5) and 2,3,4,7,8-pentachlorodibenzfuran (TEF = 0.5). On this scale 2,3,7,8-TCDF has a TEF value of 0.1.

Concern about dioxins is largely based on the very small amounts required to produce birth defects in small mammals (Table 6.22). Guinea pigs fed 2,3,7,8-TCDD in corn oil at a dose of 6 mg/kg suffered lethality of five from eight animals. One death from such a group occurred at 1 mg/kg. LD_{50} values from ingestion of polluted soil samples were about one quarter those found in corn oil.

After the Vietnamese war US veterans initiated claims against the

Table 6.21 Comparative toxicities (cf. p. 46)

Material	Probable lethal dose in humans (mg/kg)
Cacodylic acid	500–5000
Picloram	500–5000
2,4-D	50–500
2,4,5-T	50–500
DDT	50–500
Strychnine	<5
Ethanol	5000–15000

Table 6.22 Teratogenetic effects of 2,3,7,8-TCDD on mice and rats (Meyers, 1988)

	Dose (g/kg)	Form administered	Effect*
Rate	0.5	Oral	Internal bleeding
	>1.0	Sub-cutaneous	Kidney damage
Mouse	1–3	Sub-cutaneous	Kidney damage
Hamster	also susceptible at low g/kg doses		

* Determined by the ED_{50}, the dose required to produce a positive effect in 50% of the sample.

Government claiming adverse affects on their health and that of their families as a result of contact with dioxins in Agent Orange. These actions were prompted in the first instance by a Colombia documentary report broadcast on television in 1978. The situation was complicated by the ban on 2,4,5-T by the recently established (1971) EPA, although at the time there was no evidence of long-term injury to humans. There followed a long and intensive screening of groups who had been exposed to Agent Orange with control groups who had not. This work was made possible by the high sensitivity of the methods of gas chromatography (p. 62) and included studies of residual dioxin in adipose tissue and blood. The outcome was that no distinction could be drawn between the two and one such finding is shown in Figure 6.21. It is evident from this study that both the distribution pattern and the dioxin levels are essentially the same.

Carcinogenicity.   Media comment frequently implies that the dioxins are carcinogenetic to humans. There was a report in 1979 by Hardell and Sandstrom that raised the possibility of induction of relatively rare soft tissue cancers. This has subsequently been discounted by the Evatt Commission adjudicating cases in Australia and also by Sir Richard Doll.

Studies of US veterans showed that four deaths could be attributed to this condition out of 44 000 individuals who had been in Vietnam, a rate not significantly different from five deaths from this same cause out of a control group of 78 000 other veterans.

In 1984 seven manufacturers of Agent Orange eventually paid $180 million in an out-of-court settlement, but as a result there was no admission of responsibility, no case law and no Federal compensation.

The Séveso incident (10 July 1976) (Pocchiari et al., 1983).   In contrast to the use of defoliants in Vietnam, which affected an area of 1.3×10^6 hectares over a period of 7 years, the Séveso pollution was local to the ICMESA factory which then ceased production.

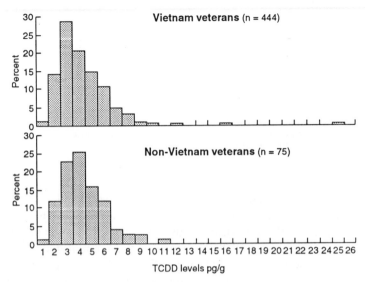

Figure 6.21   Levels in serum of 2,3,7,8-TCDD in Vietnam veterans (444) and of a control
group (75) (Lathrop, 1988).

In the first half of 1976, production of 2,4,5-trichlorophenol had risen to 142 000 kg and a typical reactor batch contained (values in kg):

Sodium trichlorophenate	2000	Sodium hydroxide	360
Ethylene glycol	1000	Sodium chloride	540
Ethylene diglycol	1300		

It was estimated that of a total load of 5200 kg there remained 2900 kg containing 600 kg of chlorophenolate. TCDD release was put in the range of 13–60 kg. The analytical detection limit of the dioxin was 0.75 $\mu g/m^2$ and it was found at a maximum limit in soil at 20×10^3 $\mu g/m^2$.

On the day, with lack of temperature control, solvents under pressure blew out the safety valve and a cloud of reagents rose to a height of 50 m and was carried on a south-easterly wind over the lightly populated area (Figure 6.22).

The background level of TCDD outside the three control zones was 1–2 ppt. Unfortunately the problem was exacerbated by a delay in enforcing control measures. Sixteen days elapsed before the population was evacuated from the highly polluted zone A and the opportunity to fix chemicals where the plume had settled on long grasses and plants was lost when a heavy rainstorm carried the material into surface soil. Ideally the procedure would be to spray the grasses with an aqueous dispersion of vinyl acetate; this sets to a polymer so trapping the pollutants, which may then be removed by harvesting the grass or other crop.

Some homes in zone A were reclaimed, floors and walls in the school

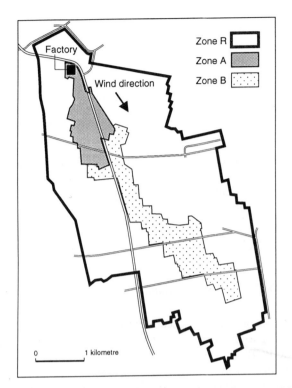

Figure 6.22 The zones affected by the ICMESA dioxin leakage. Zone A, 110 ha, population 730, TCDD in soil 400 ppt; zone B, 270 ha, subject to area specific regulation; zone R 1430 ha subject to area specific regulation (from Pocchiari *et al.*, 1983).

were cleaned daily until surface levels fell below 10 mg/m². To prevent the export of chemicals, zone B was regulated by a ban on the breeding of animals or the planting of vegetables. Children and pregnant women were required to evacuate the area by day and there was to be no procreation. Milan was 70 km to the south and there was concern to protect this densely populated area from the risk of exposure to the dioxin. The cost of the clean-up was borne by the Swiss parent company, Hofmann La Roche.

A model operation was mounted to remove and incinerate the 2500 kg of residual chemicals in the plant and over 5 t of equipment was decontaminated by Hofmann La Roche specialist workers. In a major disposal the less exposed artefacts and buildings were committed to a closely monitored landfill.

Teratogenic consequences. Great concern was felt because of the toxicity of TCDD in small mammals (pp. 238–9), but at Séveso studies showed that no significant abnormalities arose in the 5 years following the incident. There were 12 680 live births from 14 690 pregnancies – 4.9% of

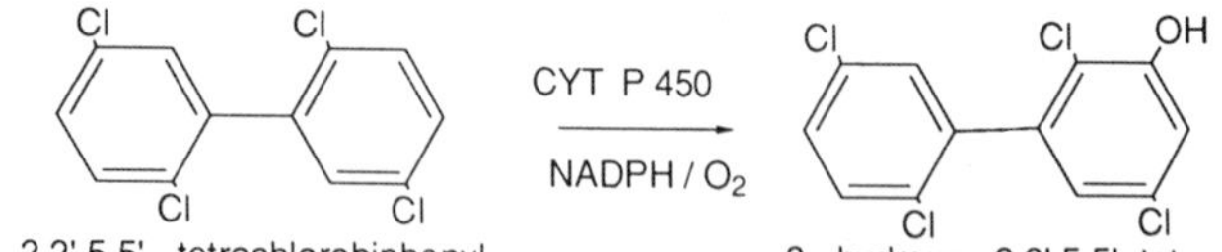

2,2',5,5' - tetrachlorobiphenyl

3 - hydroxy - 2,2',5,5'- tetrachlorobiphenyl

Figure 6.23 Direct hydroxylation of a PCB.

women of childbearing age were pregnant, which was lower than the national average but this was attributed to social rather than biological factors; there was a general decline throughout the Lombardy region. The abortion rate of 19% was high but just within a range worldwide of 10–20%, while 3.5–6.7% of children had birth defects also within the accepted range of the WHO classification.

6.4.12 Metabolism of chloroaromatic compounds

Pollutants of this group are susceptible to oxidation by P450 cytochrome enzymes leading to the formation of arene oxides and subsequent excretion (p. 210). The PCBs are subject to this reaction but also undergo oxidation by direct hydroxylation (Figure 6.23) (Preston, 1983). Electronegative chlorosubstituents reduce the rate of oxidation, which depends on abstraction of electrons, and PCBs of higher chlorine number tend to accumulate in human tissues (Borlakoglu and Dills, 1991).

6.4.13 Disposal of organochlorine compounds

These substances are largely classified as hazardous or toxic wastes and as such must be disposed of by incineration and not directly to landfill. Organochlorines are difficult to incinerate efficiently bcause of their low flammability and because of the range of reactions which can occur during combustion.

Toxic waste incinerators are operated ‘in house’ by major industrial producers or else by transport to ‘merchant incinerators’. In the UK over 50 sites are registered with the HM Inspectorate of Pollution who monitor their efficiency at the time of commissioning. Conditions for the incineration of PCBs require exposure to temperatures in the range of 1000–1200°C with oxygen in 3% excess during a residence time of 1.5–2.0 s: in the USA the EPA require an efficiency of 99.9999%. This is barely credible as it implies an accuracy of measurement of 0.0001% for such a process, which is difficult to analyse owing to the additional uncertainties of feed rate, positioning of the sampler and the extent to

Table 6.23 Estimated PCDD emissions in southern Ontario

Source	PCDD (g/year)
Industrial incinerators	65
Automobiles	>67
Wood for fuel	1260
Municipal incinerators	4522

which the products are taken up on emitted particulates. With the eyes of the world on it the incinerator ship Vulcan disposed of 10 000 t of Agent Orange with an efficiency of 99.9% which left 0.1% or 10 t unacccounted for.

In the USA the problem has been highlighted by objections by citizens groups and by state legislatures to the incineration of chemicals from disposal of chemical weapons (Silton, 1993), including organophosphorus nerve agents (p. 248).

Incomplete combustion of PCBs leads to the formation of PCDDs and PCDFs by a mechanism which is similar to that shown in Figure 6.18. In areas which are efficiently monitored, such as Southern Ontario, the normal level in air of the TCDDs is about 10^{-5} ng/m^3. The average adult inhales 20 m^3/day which at this level is $20 \times 10^{-5} \times 10^{-9} = 0.2$ pg/day; the intake from food will be greater by more than an order of magnitude. Annual emissions in such an environment have been estimated (Table 6.23) (Paterson *et al.*, 1990).

Municipal solid waste incinerators (MSW) are regarded as the principal producers of PCDD/PCDF (Oppelt, 1987). Complex reactions leading to unexpected products can occur within incinerators; for example, in the range of 6–800°C chloroform gives rise to hexachlorobenzene. Domestic waste contains aromatic precursors such as lignin and polystyrene, PVC provides a source of chlorine and synthesis of PCDDs has been shown to occur near 600°C. These lower temperatures follow the addition of too high a proportion of vegetable matter to an MSW. It was shown in a Quebec study (Finkelstein *et al.*, 1990) that unsatisfactory emissions could be brought within proper limits by changing the configuration, the air supply, and the controls: PCDD/PCDF emissions of up to 4000 ng/m^3 were reduced to 50 ng/m^3. A typical output is 30 000 m^3/h which operated inefficiently over a working year of 5000 h amounts to:

$$4 \times 10^3 \times 3 \times 10^4 \times 5 \times 10^3 \text{ ng/yr} = 600 \text{ g/yr}$$

6.4.14   Cremation or burial (Ayres, 1987)

In the UK at 1987 prices landfill cost £5/t and incineration £15/t but the latter alternative is favoured by rising land values and shortage of suitable sites. Incineration reduces the bulk of domestic waste to one-fifth, an important factor when landfill amounts to 0.5 t per head per year. Although the bulk for disposal to landfill is reduced, the fly ash from municipal incinerators is much higher in PCDD/PCDF than the raw waste, a disadvantage brought about by synthesis within the incinerator. In a test group of six municipal units the levels of TCDDs ranged between 3 and 2600 ng/g.

6.4.15   Use of decay organisms

TCDD is attacked by light and by microorganisms (cf. pp. 39–42) and has a half-life of about 1 year in soil. Attempts to dispose of chlorinated waste by microorganisms has been researched (Bennett and Olmstead, 1992), but it is difficult to divert them from their natural metabolism, also, as seen previously (p. 242), highly chlorinated compounds are resistant.

Chlorinated lignin (p. 199) and chlorophenols in the bleaching effluent from paper mills have been successfully degraded by the white rot fungus, *Phanerochaete chrysosporium* (Pellinen *et al.*, 1988).

**6.5
Natural,
organophosphorus and
carbamate pesticides**

In advanced countries the use of organochlorine pesticides has declined and some are banned unless 'no suitable safe alternative exists'. Acceptable alternatives are commonly found among the group to be discussed but unfortunately they cost about three times as much to produce, consequently in Third World countries DDT, Aldrin and Dieldrin continue in large scale use.

6.5.1   Naturally occurring pesticides

Although chemists are often accused of introducing dangerous substances into the environment very many toxic compounds are produced by plants and animals (cf. p. 45). The natural pesticides provide an example of this.

Pyrethrins.   These are obtained largely from the plant *Chrysanthemum cinerariaefolium*. The insecticidal activity of the flowers was known to the Chinese in the first century AD but commercial production of pyrethrum did not begin until 1850 and the first evidence of their structure was obtained in 1920 by Ruzicka and Staudinger. About half the current world production comes from Kenya; other major sources are Russia and Japan. In 1983 users spent 630×10^6 to treat 52×10^6 ha.

Each flower head contains 2–4 mg (*c.* 2%) of a mixture of active components and these are extracted with light petroleum. The four

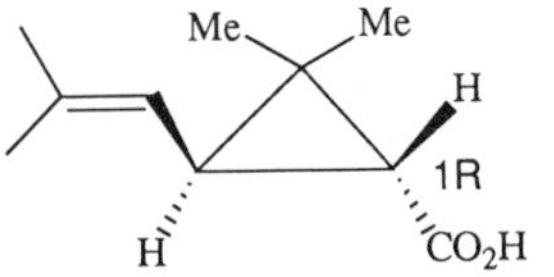

45 (1R) chrysanthemic acid

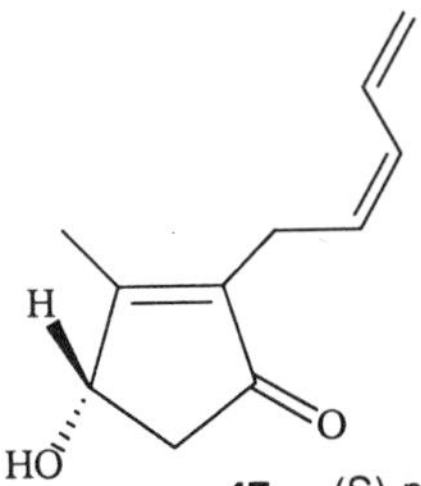

47 (S) pyrethrolone

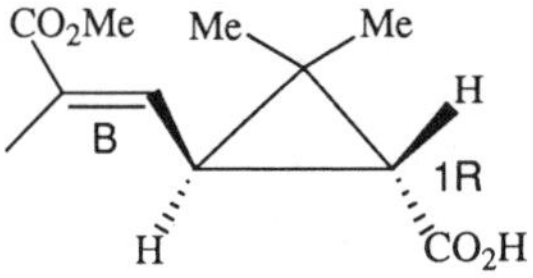

46

(1R) pyrethric acid

(+) *trans*

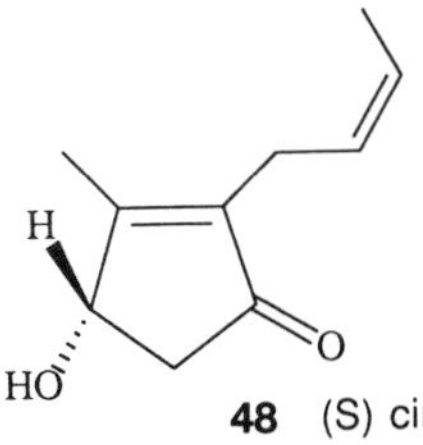

48 (S) cinerolone

Figure 6.24 Principal components of the pyrethrins.

principal compounds are esters formed by combination of either of the acids, chrysanthemic (**45**) and pyrethric (**46**), with one or other of the alcohols, pyrethrolone (**47**) and cinerolone (**47**).

Structure–activity relationships (Elliot and Janes, 1978). Key requirements are: (i) a bulky group like *gem* – dimethyl two places removed from the ester function and (ii) unsaturated centres at the two extremities of the molecule.

The activity is also related to the absolute configuration. Chrysanthemic acid, with its two asymmetric centres, can exist in four enantiomeric forms. One of these has been shown above and the remaining three are structures (**49–51**).

49

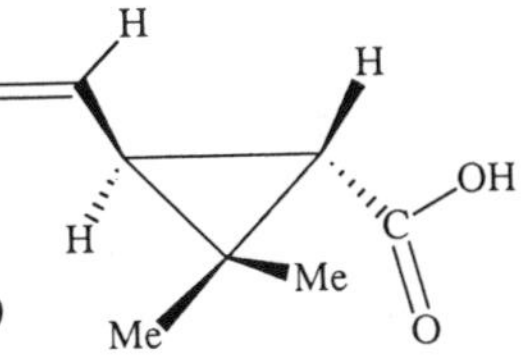

(-) *trans* (S)

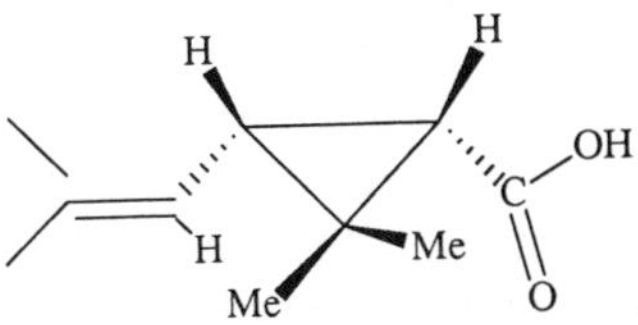

50

(-) *cis* (S)

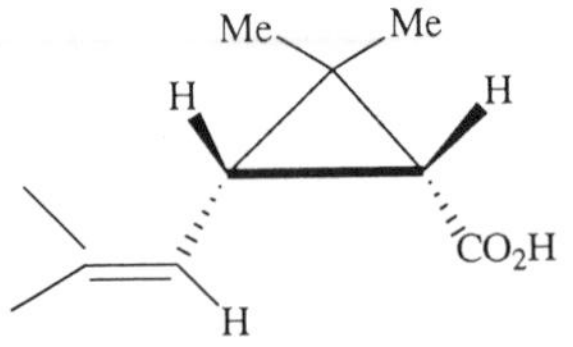

51

The *cis-* and *trans*-esters of the dextrorotatory (+) (*R*)-acids are 20–50 times as active as those of the corresponding laevorotatory (−) (*S*) isomers. Evidently when the key structural elements relate to this (*R*) configuration they can interact with enzymes of the target species.

The highly active pyrethrin 1 (**52**) is the pyrethrolyl ester of chrysanthemic acid; it meets the structural criteria for activity and has an LD_{50} of 0.3 μg/insect. The (+)-*trans* ester is slightly more toxic to mammals than the (+)-*cis* analogue with an LD_{50} of 5 mg/kg in rats by the intravenous route. The result of the interaction with the target is similar to that which follows exposure to DDT and death follows from hyperexcitation of the nervous system.

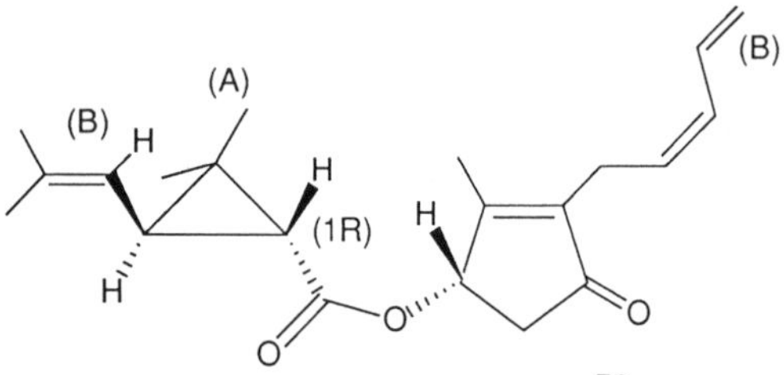

52

The advantages of the pyrethrins lie in their unsaturated aliphatic structures which undergo rapid photolytically induced oxidation, which means that they have short half-lives. However, this means that they are commonly mixed with small amounts of other pesticides, such as rotenone and resmethrin, to ensure that insects do not recover after they have been knocked down.

Synthetic analogues of pyrethrin.   These have been extensively researched and a major contribution was made by Elliot and Janes at the Rothampstead Research Station (Figure 6.25). The first of these compounds was allethrin (**53**) prepared from a racemic mixture of (*R,S*) chrysanthemic acids. In bioresmethrin (**54**) the cyclopentenone ring of the natural product is replaced by the structurally simpler benzylfuran and yet the compound is extremely potent, although still short-lived due to its photosensitivity. Some of the thinking behind the commercial development is illustrated by permethrin (**55**), which includes the easily accessible phenyl ether group and the inclusion of two chlorine atoms in the chrysanthemic acid part.

Figure 6.6.2 Some Synthetic Pyrethroids

Allethrin   **53**

(1R) form shown
0.1 µg/insect (housefly)

chrysanthemoyl :—O—CH₂

Bioresmethrin   **54**

LD₅₀ 0.005 µg/ insect

Deltamethrin   **56**

Permethrin   **55**

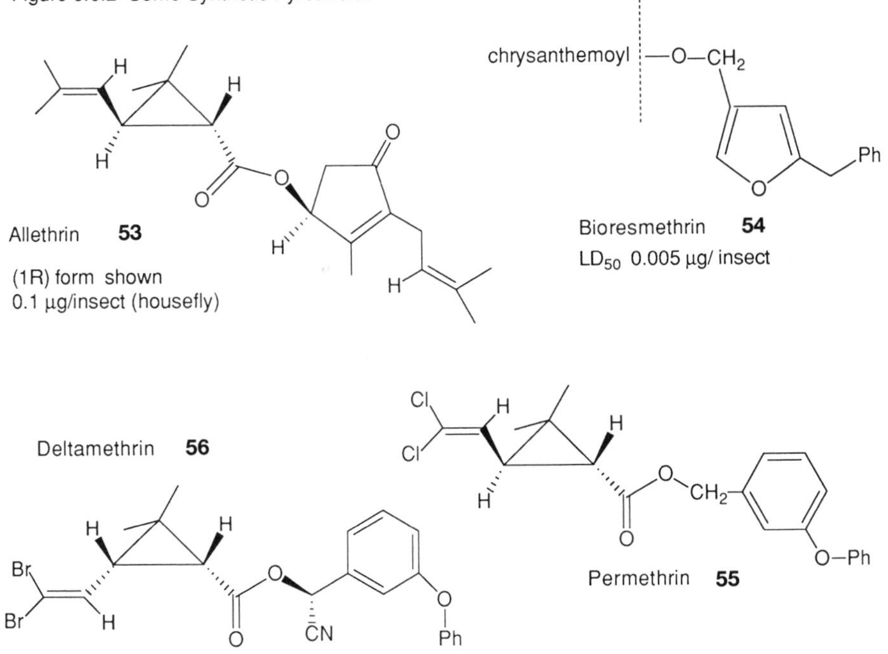

Figure 6.25   Some synthetic pyrethroids.

These electronegative substituents were seen (pp. 228, 230) to make for too long for a lifetime in polychlorocompounds; here they have the desirable effect of usefully extending the lifetime of an overly sensitive substance. Deltamethrine (**56**) is another example. The toxicity towards mammals remains low although fish are more susceptible to poisoning as shown in Table 6.24, which gives the LD_{50} values for fish and rats.

Permethrin is supplied as a mixture (R,S) with geometrical isomer ratio in the range of 40–50: 60–50; it is marketed as an insecticide under several tradenames, for example Ambush (ICI PLC), Cooper powder (Wellcome

Table 6.24   Toxicity data for some pyrethrins

Compound	Rat (mg/kg)	Fish (mg/l)	Comment
Pyrethrin	600–900[a]	114 (catfish)	Toxic to bees
Allethrin	1100	17 (trout)	
Bioresmethrin	7000	0.5 (guppy)	Toxic to bees
Permethrin[b]	400–4000	0.009 (trout)	
Deltamethrin[c]	1000	0.0012 (sunfish)	Harmless to bees outdoors

[a] Depends on sex, age and *cis/trans* ratio
[b] Half-life in soil < 38 days.
[c] Biodegrades in soil within 1–2 weeks, in plants in 10 days

Foundation Ltd), Permit (Pan Britannica Ltd), as Rentokil Musk control and in mixtures with other substances. Deltamethrin is a single isomer.

When the complexity of these structures is compared with that of DDT it can be appreciated that their synthesis is more costly than the route to DDT, depending as this does on cheap, readily available starting materials (p. 224).

Other naturally occurring pesticides. Although they are less important than the pyrethrins mention should be made of the rotenoids obtained from the roots of *Derris elliptica*, of which rotenone (57) is a typical component (Crombie, 1980). Although toxic to fish and many insects the rotenoids are not harmful to mammals; the rotenone LD_{50} to rats is 135 mg/kg. Its principal use today is in horticulture for the control of aphids, caterpillars, sawflies and wasps. The symptoms of insect poisoning by rotenoids differ from the hyperactivity associated with DDT and the pyrethrins.

Rotenone **57**

(S) (-) -Nicotine   **58**
b.p. 247° $[\alpha]_d$ 161°
sol. in water

Also of interest is nicotine, first shown to be the active principle of tobacco extracts in 1828. The naturally occurring active isomer shown is the $(-)(S)$ form. Its use has declined rapidly because of its high mammalian toxicity. The heart rate of adult humans responds to as little as 2 mg of the bitartrate, equivalent to 0.5 mg of nicotine. Serious poisonings were frequent and in extreme cases lead to death from respiratory failure; the estimated fatal dose in a human adult is about 100 mg.

Nicotine becomes bound to the red cells in the blood of smokers and reaches levels in the range of 0.4–2.0 µg/g. It is linked to heart disease and may be excreted in human milk at a level of 0.5 µg/g. The action of the molecule is critically linked to its dimensions leading to its interference with cholinesterase in a manner which is discussed on p. 249.

6.5.2   Organophosphorus pesticides (Fest and Schmidt, 1982)

Compounds of commercial value were developed from work on nerve poisons which was carried out during the Second World War by Saunders in Cambridge and Schrader in Germany. The simplest of these compounds is tetraethyl pyrophosphate (**59**), which is highly toxic to mammals as is the warfare agent Sarin (**60**).

59 TEPP   **60** Sarin   **61** Choline

The mechanism of action.   The toxic action takes place at the synapses, or junctions between nerve cells, where impulses are transmitted chemically through the acetylation of choline (**61**). When functioning normally this reaction is rapidly reversed through the action of cholinesterase, so that stimulation ceases and the site is ready to receive a new impulse. The nerve poisons act by inhibiting cholinesterase which leads to the failure of nervous transmission, convulsions and death. The action of nicotine is similar in that its critical dimensions are sufficiently close to those of acetylcholine for this to be displaced at the synapses, so disrupting the nervous system.

The simplified scheme (Figure 6.26) shows how amino acid residues on the peptidic surface of the enzyme interact in a stereo-dependent way with acetylcholine. The process can for clarity be separated into steps:

1. The positively charged end of the substrate is attracted to a carboxylate ion on the surface.
2. A specific acidic group of the enzyme forms a hydrogen bond to the acetoxy group so exposing it to nucleophilic attack.
3. The acetyl group is thereby transferred to a serine residue on the enzyme and choline is regenerated (equation 6.19).

Sarin and related phosphorus compounds interfere with the action of the enzyme by converting the serine into a type of phosphorus ester. Unlike *O*-acetylation this reaction (equation 6.20) is only slowly reversed and so the inserted group blocks the normal transmission of impulses.

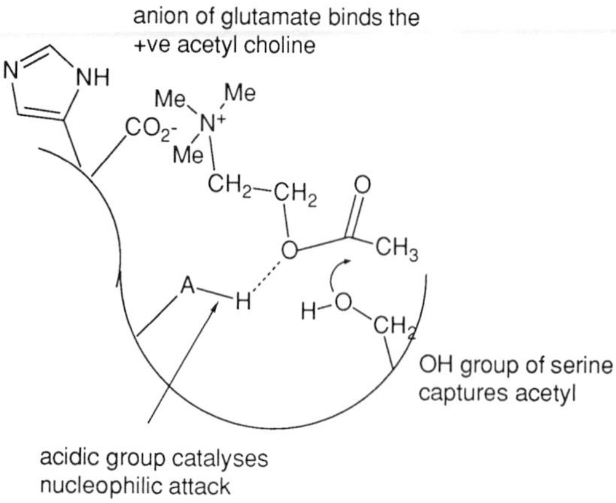

Figure 6.26

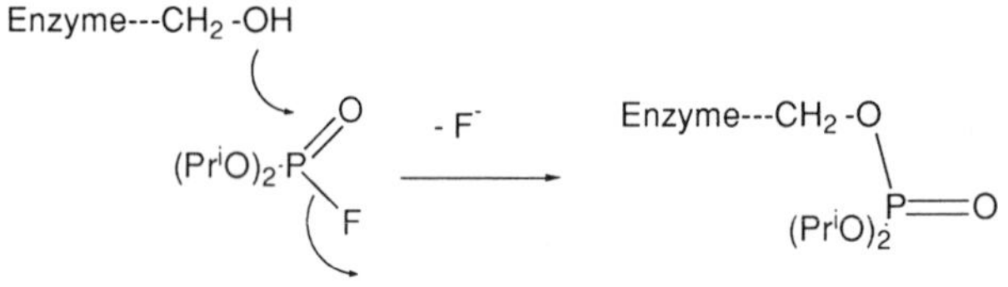

$$\text{(6.20)}$$

This section concludes with mention of carbamate pesticides. They possess an ester function and react in a very similar manner (equation 6.21) with the insertion of a blocking group.

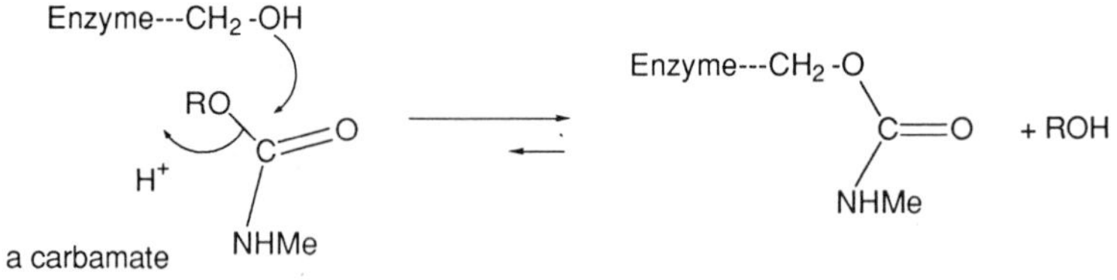

$$\text{(6.21)}$$

Some individual compounds. Those singled out for comment have been chosen to illustrate a range of properties from among those of greatest commercial importance.

Parathion was introduced by Shrader in 1952 and was one of the earliest in use in agriculture being effective against aphids and red spider mite. The methyl ester was found to be preferable as it retained the activity against pests but with less risk to mammals, however this is still considerable (Table 6.25) and Parathion is increasingly being replaced by less toxic compounds.

Parathion is synthesized by the reaction:

$$(6.22)$$

which resembles the preparation of a carboxylic ester from an alkoxide and the acid chloride. Since Parathion is an ester of thiophosphoric acid it is rapidly detoxified in the environment by hydrolysis:

$$(6.23)$$

The thioesters do not react so readily with cholinesterase and are not pesticides, but the mixed function oxidases (MFO) of insects converts them into the toxic oxyesters and so encompass their own destruction:

$$(6.24)$$

Parathion and other phosphorothioates are known as proinsecticides; they are only toxic to those animals whose MFO can effect the change into the oxyphosphate.

Malathion (Figure 6.27) was also introduced in 1952 and is less toxic to mammals than Parathion (Table 6.25). Consequently it is used for the protection of a wide range of crops including cotton, fruit, potatoes, rice, vegetables, stored grain and mammalian parasites. It is marketed under the cloak of 29 different tradenames.

Chlorpyrifos was introduced in 1965 for use in major crops such as citrus fruits, coffee, cotton, maize and also against the mosquito. The dose rate is in the range of 200–1200 g/ha and in soil it persists for only 60–120 days, being cleaved by hydrolysis into phosphate and trichloropyridine-2-ol.

Iodofenphos, is also a contact insecticide, like the above compounds, but in addition it has systemic properties. That is to say it is translocated in plants and persists long enough to make the host itself toxic to insects that eat or suck it, but not to those that merely alight on it. It was introduced in 1969 and has an important application in poultry houses.

Dichlorvos is of low molecular weight and is very volatile, which leads to its incorporation into slow release insecticidal strips for the control of flies, moths and mosquitoes. It is also effective in plants against sucking insects and as an anthelmintic. Although appreciably toxic to mammals it is not persistent being rapidly hydrolysed.

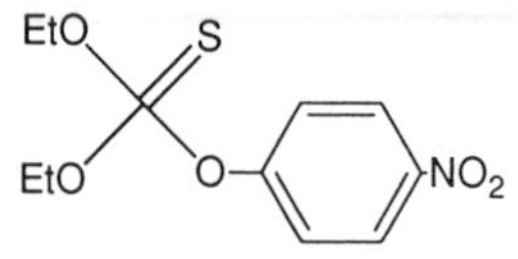

PARATHION
O,O - diethyl - O - 4 - nitrophenyl-
phosphorothioate

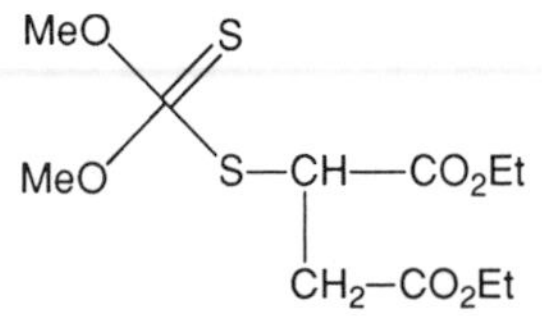

MALATHION
O,O - dimethyl - S - 1,2-(ethoxycarbonyl)-
ethyl phosphorothioate

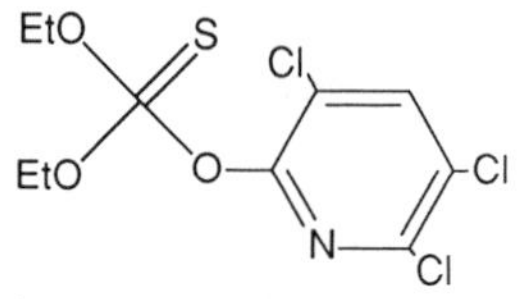

CLORPYRIFOS
O,O - diethyl - O - 3,5,6,- trichloro-2-
pyridyl phosphorothioate

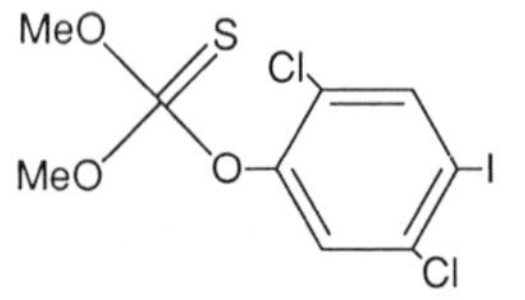

IODOFENPHOS
O,O - dimethyl - O - 2,5 - dichloro - 4 - iodo-
phenyl phosphorothioate

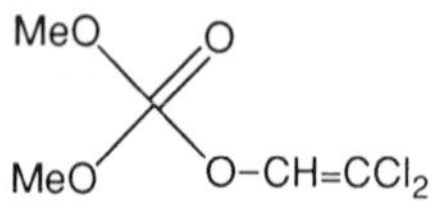

DICHLORVOS
2,2 - dichlorovinyl dimethyl phosphate

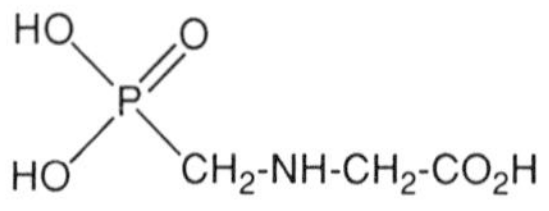

GLYPHOSATE

Figure 6.27

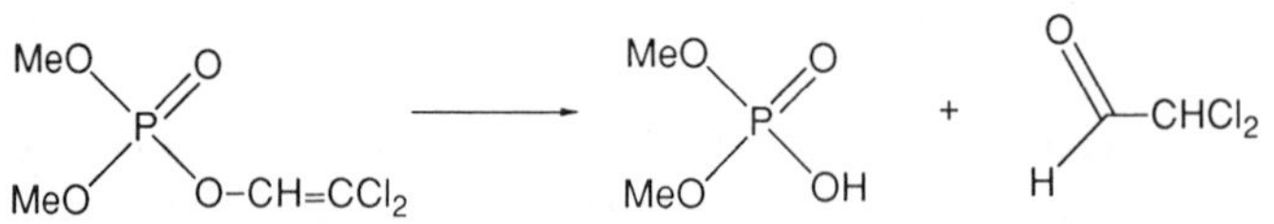

dimethyl hydrogen
phosphate

dichloroacetaldehyde

$$(6.25)$$

It is further degraded by the action of MFO to chloride ion, phosphoric acid and carbon dioxide.

Glyphosate activity was first reported in 1971. It is used as a herbicide and is widely effective at doses of 0.3 kg/ha for annual weeds and up to 2.2 kg/ha for perennials; in soil it acts as an insecticide. Common trade names are Tumbelweed, Touchdown and over 20 others.

The toxic action differs from the thiophosphate esters and depends on the fact that glyphosate is a zwitterion and a modified glycine. It acts as a

Table 6.25 Toxicity data for some organophosphorus pesticides

Compound	Oral LD$_{50}$ rat mg/kg	LC$_{50}$ fish mg/l	Persistence	Comment
Parathion (Et)	14–24	2.7 trout	2–8 days on plant, longer elsewhere	Dimethyl safer MRL[b] 0.2 fruit and vegetables
Malathion	2800[a]	0.1 bluegill	Excreted in < 24 h	MRL 8.0 cereals 0.5 fruit and vegetables
Chlorpyrifos	130–160	0.003	$t_{1/2}$ soil 60–120 day	MRL 0.3 citrus 0.2 vegetables
Iodofenphos	2100	0.1 trout	—	systemic
Dichlorvos	60–80	0.17 bluegill	$t_{1/2}$ water 20–80 h $t_{1/2}$ mammals 25 min	MRL 2.0 wheat 0.1 fruit and vegetables
Glyphosate	5600	120 sunfish	Low	Rapidly excreted by mammals

[a] The oxyphosphate, Malaoxon, has an LD$_{50}$ of only 90 mg/kg
[b] EC limit (μg/g)

glycine mimic and becomes accepted into peptides where it blocks normal development. The half-life in soil is less than 60 days.

6.5.3 Carbamate pesticides

There are currently about 40 of these compounds which have a substantial share in the market and many of them are systemic in plants. Their activity against cholinesterase (p. 249) was first suspected in the naturally occurring alkaloid physostigmine (Figure 6.28). This is, however, an ionizable base and derived ions cannot penetrate the sheath which protects the nervous system of insects. Those compounds which have come into use are therefore not ionizable and have the general structure (**69**). Notice that this resemblance to acetylcholine favours it in competition for the sites on cholinesterase (Figure 6.26).

Carbaryl is a cheap and widely used compound which was introduced as early as 1956 by the Union Carbide Company. It is synthesized by the reaction of α-naphthol with phosgene, which is essentially a dibasic acid chloride. This means that after the first esterification step an amide link can be established by reaction with the amine in the second step:

$$(6.26)$$

Carbaryl is effective against pests in cotton, fruit and vegetables and can be

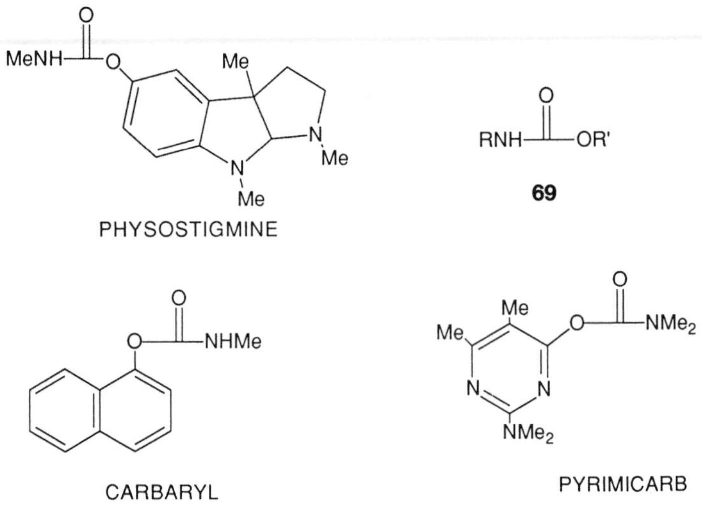

Figure 6.28 Carbamates and carbamate pesticides.

used as a substitute for DDT. The mammalian toxicity is moderate; for the rat the LD_{50} is 500 mg/kg while 200 µg/g in the diet showed no effect over 2 years.

Pirimicarb was introduced in 1968 by ICI after it was discovered that carbamates which had a pyrimidine carrier group were less toxic to mammals: for pirimicarb the oral LD_{50} in rats is about 150 mg/kg. It is more expensive than carbaryl but is effective against aphids including those which have become resistant to organophosphorus insecticides.

Mention must also be made of two triazine relatives of pyrimicarb, Atrazine and Simazine. The former is used as a pre-emergence weed killer in corn and for the culture of sugar cane and pineapples; Simazine is

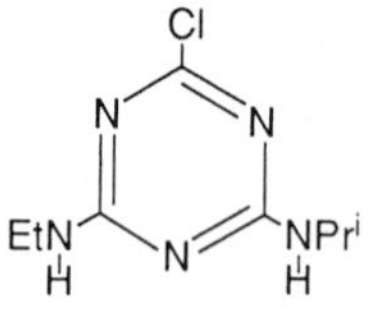

ATRAZINE
water sol.70 µg/g at 25°
m.p. 175°

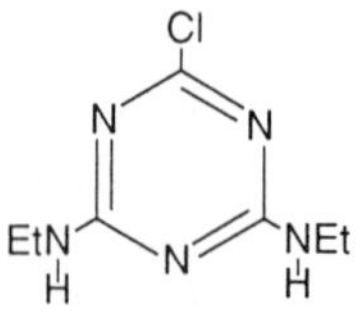

SIMAZINE
water sol. 5 µg/g at 20°
m.p.227°

effective as a herbicide. Atrazine has a half-life in soil of 30 days or less depending on type, while Simazine was reported as being completely displaced within twelve months. Unfortunately, although they are not persistent, soil organisms act on them to increase their solubility in water by replacing the chlorosubstituent with hydroxyl. These metabolites have polluted supplies in the Thames Valley and consequently both Atrazine and Simazine been placed on the red list in the UK.

For a survey of the numerous other carbamates and other specialized pesticides the reader is referred to Cremlyn (1991).

Odours play an important part in people's lives and are a very complex sensation. With regard to pollution, however, the main considerations are unpleasant odours, their sources and control. The odours which cause the most complaints are mainly associated with livestock production, industries rendering animal wastes, waste disposal (especially sewage sludge and animal manure disposal onto land) and some industrial processes (Elsom, 1987; Clarke, 1992). However, even 'pleasant' odours such as those from perfumeries can be a nuisance when present in excessive quantities, in the wrong place at the wrong time. Although odours are usually a local problem, they can also be carried several kilometres downwind and therefore offend a relatively large number of people. The 'Bedford smell' (mercaptans and SO_x) from the brick kilns near the town of Bedford is a good example of this (Owen, 1981) (see section 5.4.3).

6.6 Odours

In the UK in 1989/90, there were 9800 official complaints of smell from industrial operations and 3800 related to agricultural practices (e.g. manure spreading) (Murley, 1992). In the US, 10% of the population (25 million people) considered odours a problem which required some form of control (Elsom, 1987).

6.6.1 Important properties of odours.

1. Intensity – the strength of the odour sensation which depends on the concentrations of odour substances present.
2. Odour character – properties which enable substances to be detected (those odours with a recognizable character are considered to be the most annoying).
3. Hedonic tone – scale of pleasantness and unpleasantness, in the context of a specific situation;
4. Frequency/duration of odour releases – important determinants of nuisance.
(Murley, 1992).

It is important to note that many odours have very low threshold concentrations at which they can be detected, for example alkyl mercaptan, which has a disagreeable garlic-like odour, has a threshold of 0.00005 mg/1 (or 0.05 ppb). In contrast, SO_2 has a threshold of 0.009 mg/1 and NH_3 a threshold of 0.37 mg/1 (Cheremisinoff, 1991). Odours are usually mixtures of gaseous substances which stimulate the olfactory senses. They are often present at low concentrations near to limit of detection and are therefore difficult to analyse accurately (GC-MS is the most suitable chemical method). One way of measuring an odour is to determine the number of times that it needs to be diluted with clean air until 50% of the people on a test panel can no longer detect it (Clarke, 1992).

Odour fatigue can also occur when a subject becomes acclimatized to an odour and no longer notices it. This can be dangerous in the case of toxic gases such as H_2S because their odour provides a warning of the presence of a toxicity hazard.

Some of the important malodorous substances caused by sewage treatment and other major causes of odour nuisance include hydrogen sulphide (H_2S), skatole (C_9H_9N), ethyl mercaptan (C_2H_6S) and methyl mercaptan (CH_4S), cadavarine, and various amines. In the case of sewers, the major odour problems occur when the sewage does not move fast enough and anaerobic conditions develop. This is often worst where garbage grinders are used which add a lot of putrescible solid material to the sewage. The most unpleasant odours from industrial processes are those from substances containing sulphur and nitrogen (mercaptans and amines, respectively) (Elsom, 1987).

6.6.2 Methods of odour control

1. Process and plant modifications.
2. Dilution by the atmosphere.
3. Odour modification (adding other chemicals to alter the response to the original odour) (masking or counteracting), e.g. air wicks or aerosol fragrances for indoor use. Masking is controversial because it involves adding more chemicals to the atmosphere and therefore is the opposite of cleaning-up.

(Murley, 1992).

6.6.3 Methods of odour treatment

1. Incineration by raising the emission temperature of a process sufficiently and for long enough to completely oxidize the malodorous organic compounds in the emissions. This can be done by using thermal afterburners but the fuel cost can be prohibitive unless heat recovery, catalysts, or another furnace is used. Temperatures need to be at least 650–800°C but can be lower (650°C) with catalysts and good mixing with air to ensure complete oxidation.
2. Absorption of malodorous compounds in reagent solutions (commonly water).
3. Adsorption on activated carbon (commonly used in domestic and catering kitchens).
4. Chemical oxidation with ozone, or hydrogen peroxide.

5. Biological treatment – passing the odorous air through columns with microorganisms growing on the packing or through peat or through soil. (Murley, 1992; Clarke, 1992).

Alawi, M. A. and Heidman, W. A. (1991) Analysis of PCBs in environmental samples from the Jordan valley. *Toxicol. Environ. Chem.*, **33**, 93–99.

Ayres, D. C. (1987) Cremation or burial. *Chem. in Britain*, **23**, 41–44.

Bacon, C. E., Jarman, W. M. and Costa, D. P. (1992) Organochlorine and PCB levels in pinniped milk from the Arctic, the Antarctic, California and Australia. *Chemosphere*, **24**, 779–791.

Ballschmiter, K. and Wittlinger, R. (1991) Interhemisphere exchange of hexachlorohexanes, hexachlorobenzene, PCBs and DDT in the lower troposphere. *Environ. Sci. Technol.*, **25**, 1103–1111.

Bayer, C. W. and Black, M. S. (1987) Capilliary chromatographic analysis of VOCs in the indoor environment. *J. Chromatog. Sci.*, **25**, 60–63.

Bennett, G. F. and Olmstead, K. P. (1992) Micro-organisms get to work. *Chem. in Britain*, **28**, 133–137.

Berg, H., Kiibus, M. and Kautsky, N. (1992) DDT and other insecticides in the Lake Kariba ecosystem, Zimbabwe. *Ambio*, **21**, 444–450.

Borlakoglu, J. T. and Dills, R. R. (1991) PCBs in human tissue. *Chem. in Britain*, **27**, 815–818.

B. P. Statistical Review of World Energy (1991), British Petroleum, London.

Cheremisinoff, P. N. (1992) *Industrial Odour Control*, Butterworth-Heineman, London.

Clarke, A. G. (1992) Chapter 2 in Harrison, R. M. (ed) *Understanding Our Environment* (2nd edition) Royal Society of Chemistry, Cambridge.

Cremlyn, R. J. (1991) *Agrochemicals, Preparation and Mode of Action*, Wiley, Chichester.

Crombie, L. (1980) Chemistry and biosynthesis of natural pyrethrins. *Pesticide Sci.*, **11**, 102–118.

Davidsohn, A. S. and Milwidsky, B. (1987) *Synthetic Detergents* (7th edition), Longman, Harlow.

Elliot, M. and Janes, N. F. (1978) Synthetic pyrethroids – a new class of insecticide. *Chemical Society Reviews*, **7**, 473.

Elsom, D. (1987) *Atmospheric Pollution*, Blackwell, Oxford.

EUROSTAT (1992) *Europe in Figures*, Office of Official Publications, Luxembourg.

Fest, C. and Schmidt, K. J. (1982) *Chemistry of Organophosphorus Pesticides* (2nd edition), Springer, Berlin.

Finkelstein, A., Klicius, R. and Hay, D. (1990) The national incinerator testing and evaluation programme: air pollution control technology assessment results, in Clement, R. and Kagel, R. (eds), *Emissions From Combustion Processes*, Lewis, Chelsea, Michigan.

Glowa, J. R. (1990) Behavioural toxicology of solvents. *Drug Development Res.*, **20**, 411–428.

Goring, C. A. I. and Hamaker, J. W. (1972) *Organic Chemicals in the Soil Environment*, Dekker, New York.

Harrison, R. (ed) (1990) *Pollution Causes, Effects and Control*, Royal Society of Chemistry, Cambridge.

Harrison, R. (ed) (1992) *Understanding Our Environment*, Royal Society of Chemistry, Cambridge.

Harvey, R. G. (ed) (1985) PAH and carcinogenesis. *Amer. Chem. Soc. Symp. Series*, **283**, 1985.

Hutson, D. H. and Roberts, D. R. (1987) *Progr. in Pesticide Biochem. Toxicol.*, **6**.

International Agency for Research on Cancer (IARC) (1983) *PAH. Part 1. Chemical, Environmental and Experimental Data*, 32, WHO, Lyons.

International Agency for Research on Cancer (IARC) (1986) *Evaluation of the Carcinogenic Risk of Chemicals to Humans*, Vol. 38, Tobacco smoking, WHO, Lyons.

Jerina, D. M. and Daly, J. W. (1974) 'Arene oxides: a new aspect of drug metabolism. *Science*, **185**, 573–582.

Kamrin, M. A. and Rodgers, P. W. (1985) *Dioxins in the Environment: Hemisphere*, McGraw-Hill, Washington, DC.

Karsa, D. R., Goode, J. M. and Donnelly, P. J. (eds) (1991) *Surfactants Applications Directory*, Blackie, Glasgow.

Kennaway, E. (1955) The identification of a carcinogenic compound in coal tar. *Brit. Med. J.*, **2**, 749–752.

Kirk-Othmer (1983) *Encyclopedia of Chemical Technology*, (3rd edition), Wiley, New York.

Kleopfer, R. D. and Kirchmer, C. J. (1985) Quality assurance plan for 2,3,7,8-tetrachlorodibenzo-*p*-dioxin monitoring in Missouri, in Keith, L. H., Rappe, C. and Choudhary, G. (eds), *Chlorinated Dioxins and Dibenzofurans in the Total Environment*, Butterworth, Boston.

Kuel, D. W., Haebler, R. and Potter, C. (1991) Chemical residues in dolphins from the US Atlantic coast including bottlenose obtained during the 1987/8 mass mortality. *Chemosphere*, **22**, 1071–1084.

Lathrop, G. D. (1988) in Young, A. L. and Reggiani, G. M. (eds), *Agent Orange and its Associated Dioxin*, Elsevier, Amsterdam.

Lewis R. J. (ed) (1992) *Sax's Dangerous Properties of Industrial Materials* (8th edition), Van Nostrand, New York.

Martin, H. and Woodcock, D. (1983) *The Scientific Principles of Crop Production* (7th edition), Arnold, London.

Martin, D. E. (1990) The environment effects of oil and gas production, in Dunderdale, J. (ed) *Energy and the Environment,* Royal Society of Chemistry, Cambridge.

Meyers, V. K. (ed) (1988) *Teratogens: Chemicals which Cause Birth Defects*, Elsevier, Amsterdam.

Montreal Protocol Assessment (1991) *Report of the Solvents, Coatings and Adhesives Technical Options Committee.*

Moriarty, F. (1988) *Ecotoxicology: the Study of Pollutants in Ecosystems*, Academic Press, London.

Murley, L. (ed) (1992) *1992 Pollution Handbook*, National Society for Clean Air and Environmental Protection, Brighton.

National Academic Press (1981) *Indoor Pollutants*, Washington DC.

Oppelt, T. E. (1987) *J. Air Pollution Control Assoc.*, **37**, 558–586.

Osborne, M. O. and Crosby, N. T. (1987) *Benzopyrenes*; Cambridge University Press, Cambridge, p. 198.

Owen, K. (1981) *The Times* (London), January 23.

Paterson, S., Shiu, W. Y., Mackay, D., and Phyper, J. D. (1990) Dioxins from combustion processes: environment fate and exposure, in Clement, R. and Kagel, R. (eds), *Emissions from Combustion Processes*, Lewis, Boca Raton.

Pellinen, J., Yin, C. F., Joyce, T. W. and Chang, H. M. (1988) Treatment of chlorino bleaching effluent using a white-rot fungus. *J. Biotechnol.*, **8**, 67–75.

Pocchiari, F., Di Domenico, A., and Silano, V. (1983) Environmental impact of the accidental release of tetrachlorodibenzo-*p*-dioxin (TCDD) at Séveso (Italy), in Coulston, F. and Pocchiari, F. (eds), *Accidental Exposure to Dioxins*, Academic Press, New York.

Pottier, A. C. W., Taylor, J. C., Norman, C. L., Meyer, L. C., Anderson, H. R., and Ramsey, J. D. (1992) *Trends in Deaths Associated with Abuse of Volatile Substances 1971–1990*, St Georges Hospital, London.

Preston, B. D., Miller, J. A., and Miller, E. C. (1983) Non-arene oxide aromatic ring hydroxylation of 2,2′, 5,5′-tetrachlorobiphenyl. *J. Biol. Chem.*, **258**, 8304–8311.

Rodricks, J. V. (1985) *Calculated Risks*, Cambridge University Press, Cambridge.

Royal Commission on Environmental Pollution (1985) 11th Report *Managing Waste: the duty of care*, HMSO, London.

Schechter, A., Tiernan, T. O., Taylor, M. L., Van Ness, G. F., Garett, J. H., Wagel, D. J., Gitlitz, G. and Bogdasarian, M. (1985) Biological markers after exposure to polychlorinated dibenzo-*p*-dioxins, dibenzofurans, biphenyls and biphenylenes. Part 1: Findings using fat biopsies to estimate exposure, in Keith, L. H. Rappe, C. and Choudhary, G. (eds), *Chlorinated Dioxins and Dibenzofurans in the Total Environment*, Butterworth, Boston.

Silton, T. (1993) Out of the frying pan: chemical weapons incineration in the USA. *Ecologist*, **23**, 18–25.

Sues, M. J. (1976) *Sci. Total Environ.*, **6**, 239.

Timberlake, L. and L. Thomas (1990) *When the bough breaks*, Earthscan Publications, London.

Verschueren, K. (1983) *Handbook of Environmental Data of Organic Chemicals* (2nd edition), Van Nostrand, New York.
Wayne, R. P. (1991) *Chemistry of Atmospheres,* Oxford University Press, Oxford.
WHO (1987) *Air Quality Guidelines for Europe*, WHO, Copenhagen.
WHO (1989) *Aldrin and Dieldrin*, WHO, Copenhagen.
Young A. L. and Reggiani, G. M. (1988) *Agent Orange and its Associated Dioxin*, Elsevier, Amsterdam.

Hydrocarbons

Further reading

Coughtrey, P. J., Martin, M. H. and Unsworth, M. H. (eds) (1987) *Pollutant Transport and Fate*, Oxford University Press, Oxford.
Ferguson, J. E. (1982) *Inorganic Chemistry and the Earth*, Pergamon, Oxford.
Henderson, B. and Sellers, B. (1984) *Pollution of Our Atmosphere*, Adam Hilger, Bristol.
Myerson, J. (1991) *Mad Car Disease*, Greenpeace, London, *Pollution Handbook* (1991) National Society for Clean Air.
Vo-Dinh, T. (ed) (1989) *Chemical Analysis of PAH*, Wiley, New York, p. 8.
Wolf, S. M. (1988) *Pollution Law Handbook*, Quorum, New York.

Organic solvents

Brandrop, J. and Immergut, E. H. (eds) (1989) *Polymer Handbook* (3rd edition), Wiley, New York.
Fawell, J. K. and Hunt, S. (1988) *Environmental Toxicology, Organic Pollutants,* Ellis Horwood, Chichester.
International Agency for Research on Cancer (IARC) (1989) *Some Organic Solvents, Resin Monomers, Pigments and Occupational Exposures in Paint Manufacture*, Vol. 47, Lyon.
Russell, J. (1993) Butane abuse: the neglected killer. *New Scientist,* **1859**, 21–23.
Sheftel, V. O. (1990) *Toxic Properties of Polymers and Additives*, RAPRA, Shrewsbury.
Snyder, R. (ed) (1987) *Ethel Browning's Toxicity and Metabolism of Industrial Solvents, Vol. 1. Hydrocarbons*, Elsevier, Amsterdam.

Organochlorines

Brown, P. (1992) Who cares about malaria? *New Scientist,* **1845**, 37–41.
British Crop Protection Council (1986) Biotechnology, Crop Improvement and Production, *Monograph*, **34**.
British Medical Association Guide to Pesticides, Chemicals and Health (1992), Arnold, London.
Hayes, W. J. and Laws, E. R. (1991) *Handbook of Pesticide Toxicology*, Academic Press, San Diego.
Ragsdale, N. N. and Menzer, R. E. (1989) Carcenogenicity and pesticides, *Amer. Chem. Soc., Symposium Series*, **414**, Washington DC.

Organophosphorus and other pesticides

Kidd, H. and James, D. R. (1991) *The Pesticide Index* (2nd edition), Royal Society of Chemistry, Cambridge.
Matsumura, F. (1985) *Toxicology of Pesticides* (2nd edition), Plenum, New York.
Ministry of Agriculture, Fisheries and Food (1990) *Pesticides*, HMSO, London.
Watterson, A. (1988) *Pesticide Users Health and Safety Handbook*, Gower, Aldershot.

Part Three

Wastes and Other Multi-pollutant Situations

Wastes and their disposal 7

7.1
Introduction

Wastes and their disposal are the cause of a great deal of environmental pollution. The pollutants involved come from a wide range of sources including 150 year-old heaps of Pb–Zn mine waste in remote upland areas, highly toxic wastes from the chemical industry and nuclear power stations and household waste. Industrial manufacturing has expanded over the last hundred years and since the Second World War there has been a pronounced trend towards increasingly complex products either containing highly toxic chemicals, such as PCBs in electrical equipment, or involving toxic materials in their manufacture. With increasing affluence in the more technologically developed countries more municipal solid waste and waste water are being produced and these need treatment and disposal. Even in the home there is a much wider range of chemicals in everyday use (see also section 6.3 and chapter 2 on indoor pollution) as well as many more people pursuing hobbies which involve the use and disposal of chemicals (car engine oil, decorating and DIY products, photography, drycell batteries, etc.).

Industrial wastes, especially hazardous wastes, can contain a wide range of chemicals some of which may be unstable and others which are highly stable and difficult to denature in order to reduce their hazard to organisms or structures. Wastes can include many of the groups of pollutant substances featured in this book. The main considerations in this chapter are:

(i)   the amounts of waste produced;
(ii)   the nature of the wastes;
(iii)   the options available for treatment or disposal;
(iv)   the environmental safety of these wastes management options.

7.2
Amounts of waste produced

In 1990, OECD countries produced 9×10^6 t municipal waste; 1.5×10^9 t industrial wastes (including 300×10^6 t hazardous wastes); 7×10^9 t other wastes (including fossil fuel ash from electricity generation, agricultural wastes, mine wastes, demolition debris and sewage sludge) (OECD, 1991).

7.2.1 Industrial wastes

The OECD group, which includes the 24 most technologically advanced countries in the world, together produce 68% of the world's industrial waste and 89.6% of the hazardous and special wastes. North America alone produces 57% of the OECD's industrial waste and 54% of its municipal waste.

Total OECD industrial waste $= 1.43 \times 10^9$ t
Total world industrial waste $= 2.1 \times 10^9$ t

Eastern Europe industrial waste $= 0.52 \times 10^9$ t
Eastern Europe hazardous and special waste $= 19 \times 10^6$ t
(OECD, 1991).

7.2.2 Municipal wastes

These comprise the waste from domestic houses, offices and commercial properties but not industrial or hazardous wastes. In the USA, municipal wastes on average comprise: 41–50% paper; 9–18% glass; 10–18% yard (garden) waste; 10–17% miscellaneous; 9% metals; 8–12% food waste; and 7% plastics.

7.3
Methods of disposal of municipal wastes

7.3.1 Landfilling

This basically involves placing the waste in compacted layers in a lined pit or a mound (sanitary landfills) with appropriate leachate and landfill gas (mainly CH_4) control. The percentage of municipal wastes disposed of by landfill in selected countries is:

USA   60%
EC   70%
Japan   38%

Landfilling is not strictly a disposal method, but really one of containment or indefinite waste storage. Many landfills are used for the disposal of

Table 7.1 Municipal waste produced per capita in different countries (kg/cap) (OECD, 1991)

	mid 1970s	1980	mid 1980s	late 1980s
N. America	633	687	734	826
Japan	341	355	344	394
OECD (Europe)	277	323	346	336
OECD (all countries)	407	436	493	513

both municipal and hazardous waste. The latter are 'co-disposed' with the municipal waste utilizing the absorptive properties of the bulky municipal waste and the microbial degradation reactions which occur wherever putrescible material is buried. There is now a move towards exploiting the microbial methanogenic processes in landfills and using them as bioreactors to both dispose of waste and also produce methane which can be recovered and used as a source of energy.

The advantages of landfilling are that it is relatively inexpensive; methane production can be exploited from municipal wastes and the wastes can be recycled or treated at a later date. One particular benefit is that in some countries, such as the UK, landfilled municipal wastes can be used for the co-disposal of some types of hazardous wastes. The disadvantages are that there is a danger of leakage and pollution of groundwater, a danger of explosions from methane and the problem that there are fewer and fewer suitable sites available. Modern landfill sites need to be located in impermeable strata or sealed with an appropriate membrane before the wastes are deposited and should have a system for managing the leachate. Older landfills worked more on the principle of waste attenuation where it was accepted that leachate would leak from the site but it was intended that the rate and volumes of leachate involved would not cause any acute environmental problems. It is now realized that this type of waste disposal is responsible for many acquifers being significantly polluted with a wide range of xenobiotic and other chemicals.

7.3.2   Incineration

This involves passing the waste through a chamber at a high temperature (preferably around 1200°C) with an adequate supply of oxygen to oxidize all organic material (p. 242). However, at present, many municipal waste incinerators do not satisfy the new emission regulations mainly because they do not reach a sufficiently high temperature and do not have flue gas cleaning facilities (electrofiltration, mechanical filtration). Some incinerators are constructed to allow the recovery of some of the heat of combustion which is either used for generating steam and electricity or for neighbourhood heating.

Although incineration is regarded as a waste disposal method, it is actually only a system of waste reduction, but it does have the desirable effect of decomposing all organic compounds in the process. An ash of inorganic residues is produced which requires disposal in a controlled landfill. However, the disposal of incinerator ash is causing some political debate, because if it is classed as a hazardous waste its disposal will be much more expensive than ordinary municipal waste, which, together with the relatively high cost of the incineration operation itself make the economics of this waste management option look less favourable.

More than 50% of the municipal waste in Japan, Sweden and Switzerland is incinerated, and more than 20% incinerated in Belgium, France, Germany, Italy, Luxembourg and the Netherlands. For the EC countries as a whole, of those incinerators which have been are fitted with flue gas treatment, 35% have mechanical filters, 36% have electrofiltration and just 7% have both of these combined (OECD, 1991).

The advantages of incineration are that it can dispose of 99.999% of organic wastes (including chlorinated organic xenobiotics) if properly carried out (high temperature 1200°C, adequate oxygen, etc.), it reduces the volume of waste and energy can be recovered from the process and utilized for electricity generation and combined heat and power. The disadvantages are that it is relatively expensive, there is a danger of highly toxic pollutants (PCDDs and PCDFs) being synthesized and emitted into the atmosphere if conditions are not optimal and, finally, that there is an ash produced which needs careful disposal in sanitary landfills.

7.3.3 Composting

Involves shredding and separating the putrescible fraction of municipal waste, often mixing it with other organic materials (including sewage sludge) and allowing microbiological decomposition reactions to take place. This involves regular turning of the compost in order to promote aerobic decomposition processes and allowing the compost temperature to rise sufficiently to kill off pathogenic organisms and weed seeds and other undesirable constituents. The finished compost is intended for use as a growing medium for plants, especially as a substitute for peat which is a scarce resource from an ecosystem at risk of destruction with the accompanying problems of extinction of species. Within Europe, the composting of municipal wastes is carried out to a significant extent in Austria, Belgium, France, Italy, the Netherlands, Portugal and Spain (OECD, 1991).

7.3.4 Recycling

Where materials are recycled (preferably on-site) there will be a saving in waste disposal costs as well as in inputs to the process. In addition to materials, energy can also be recovered from the incineration of combustible wastes, or the production of methane from putrescible wastes, which can make a positive contribution to energy budgets. Some examples of recycling in practice are:

- Scrap metal is used widely in steel manufacture.
- Recycled glass is being collected (bottle banks) and used on an increasing scale.

Table 7.2 Total wastes arisings in England and Wales (House of Commons Env. Comm., 1989; BMA, *Hazardous Waste and Human Health* (1991), by permission of Oxford University Press)

Waste type	Quantity (10^6 t/yr)
Liquid industrial effluent	2000
Agricultural wastes	250
Mines and quarries	130
Industrial	50
(hazardous and special	3.9)
(special	1.5)
Domestic and trade	28
Sewage sludge	24
Power station ash	14
Blast furnace slag	6
Building	3
Medical wastes	0.15
Total	2505.15

- Non-ferrous metals including Cu, Pb and Zn are recycled.
- Old engine oils are treated and reused.
- Plastics will soon be recycled on an increasing scale after a considerable research and development effort, mainly in the USA.

An example of the waste arisings in a technologically advanced country is provided by the data for England and Wales in Table 7.2.

7.4
Sewage treatment

The treatment of waste water is of vital importance for its roles in preventing the spread of infectious diseases, such as cholera (see section 6.4), and protecting river water quality and avoiding noxious odours. In England and Wales, 23×10^6 m^3 of domestic waste water and 14.1×10^6 m^3 of industrial waste water are discharged to the sewers every day (Lester, 1990). This sewage is treated by a total of 5000 sewage works around England and Wales which serve around 44×10^6 people. In addition, the sewage from a further 6×10^6 people living near the coast is discharged into the sea without treatment and up to 2×10^6 people are not connected to the sewage system and use their own septic tank systems (Lester, 1990). The discharge of sewage into the sea around the coastline of Britain has attracted a lot of attention and adverse publicity with regard to both the concern about the highly polluted state of the North Sea and its marine ecosystem and also to its possible effects on the health of bathers and surfers using the beaches. As a result of a draft EC Directive in 1989 which was accepted in principle in the UK in 1990, the direct discharge of untreated sewage to the sea will be discontinued by 1998 but there will be a delay while extra sewage works are built to handle the additional sewage.

In the UK, the original object of sewage treatment was to avoid disease and unpleasant odours from rivers carrying untreated sewage. This is exemplified by the 'Year of the Great Stink' in London in 1858 when a combination of a long dry summer with reduced flows in the River Thames and high rates of sewage discharge into the river (from an increasing number of houses with flush sanitation) caused such a bad odour problem that the business of the Houses of Parliament (which overlook the Thames) had to be adjourned for several days. There had already been three major epidemics of cholera in London before this (1831–32, 1848–9 and 1853–4). Eventually, after further cholera epidemics and the continuing odour problem a Royal Commission on Sewage Disposal was appointed in 1881. This body was instrumental in putting various remedial measures into action which resulted in marked improvements in the smell and quality of the Thames (Ellis, 1989).

The sewage treatment processes were shown to be very effective in cleaning up waste waters and so long as the concentrations of the polluting materials are reduced to levels within the capacity of the equatic ecosystems receiving the purified waste water, further purification and dilution will occur. With increasing populations in many parts of Britain, especially in London and the Thames Valley, it became necessary to re-use treated water further downstream and nowadays about 30% of all drinking water supplies involve indirect re-use, which is a much greater proportion than in most other countries (Lester, 1990).

Sewage treatment involves three basic processes (Fish, 1992):

1. the removal of polluting matter from the sewage flow as solids or slurries of solids in water (sludges);
2. the removal of polluting matter from the sewage flow and separated sludges by accelerated natural processes of biochemical breakdown brought about by microorganisms;
3. the separation of water from sludges to remove the volume of sludge for disposal.

These three processes occur during six stages in the treatment of sewage which are (Fish, 1992):

1. a preliminary screening to remove grit and solids;
2. a primary settlement stage in tanks to allow the removal of solids and grease;
3. secondary treatment involving the microbial oxidation of organic matter and ammonia and the further removal of solids;
4. a polishing treatment using sand filtration to remove very fine solids;
5. tertiary treatment under anaerobic conditions to bring about denitrification and chemical precipitation and remove N and P;
6. a sludge treatment stage involving combinations of digestion, thickening, dewatering and drying to prepare the sludge for disposal and also to produce CH_4 (which is utilized to power pumps in the works).

It is important to note that if the sewage is heavily polluted with toxic substances, many of the microorganisms involved in the biochemical processes may be killed and the treatment will be incomplete. This is soon detected and can be coped with but it causes a lot of extra problems at the sewage works.

Domestic sewage contains approximately 0.1% (1000 mg/l) of impurities about half of which are dissolved and the rest in suspension. About 70% of the impurities are organic and include proteins and urea, sugars, starches and cellulose, soap, cooking oil and greases. The inorganics include chloride, and metallic salts and road grit where storm water is combined in the sewage. This raw sewage normally has a suspended solids load of approximately 400 mg/l and a BOD of 300 mg/1 (Lester, 1990). After full primary and secondary treatment and 6 hours aeration in a diffused air sludge activation plant the final effluent should contain less than 15 mg/l suspended solids, about 15 mg/l of 5 day BOD and ammonia nitrogen content of less than 10 mg/l (Fish, 1992) (see section 2.5.1). This is discharged into rivers where it becomes mixed and diluted and could undergo further biochemical purification.

Raw water extracted for public water supply is usually disinfected with chlorine and/or ozone or UV light. There has been concern about the synthesis or organohalides in water as a result of chlorination. Recent studies in the USA have added further weight to the apparent association of a small increase in bladder cancer incidence in long-term consumers of chlorinated water. However, it is acknowledged that the possible slight risk to health is not as great as the consumption of non-disinfected water (Neutra and Ostro, 1992). However, there is obviously a need for further careful investigation of this subject since a very large proportion of the populations of technologically advanced countries use chlorine for water disinfection.

The sludge produced from sewage treatment needs disposal and at present the main disposal options are land application, dumping at sea, landfilling and incineration. Dumping at sea is decreasing and in the UK land application accounts for around 67% of sludge disposal. This currently only affects 2% of agricultural land in the UK but this land is often of good quality in the vicinity of towns (near to sewage works) and the proportion of sludge applied to land will increase as sea dumping is phased out. Sewage sludge has several beneficial properties for use on agricultural soils. It contains useful amounts of nitrogen and phosphorus which are usually applied in fertilizers and therefore the sludge has positive fertilizer value as well as being a source of organic colloids which make a beneficial contribution to soil aggregate stability and soil structure. However, it is the pollutants in sludge which limit its usefulness. Heavy metals are a particular problem because they are concentrated in the sludge and most will accumulate in the soil with continued applications of sludge (see chapter 2 and section 5.3). The maximum permissible concentrations of metals in soil after the application of sewage sludge to agricultural land are given in Table 7.3.

Table 7.3 The EC maximum permissible concentrations of heavy metals and other elements in sewage sludge amended soils (EC 1986; MAFF, 1992)

Element	Maximum concentration (mg/kg dry solids)	Maximum rate of addition over 10 year period (kg/ha)
Cd	3	0.15
Cr	400 (provisional)	15 (provisional)
Cu	100	7.5
Hg	1	0.1
Ni	50	3
Pb	300	15
Zn	300	15
Mo*	4	0.2
Se*	3	0.15
As*	50	0.7
F*	500	20

*Elements not subject to EC Directive 86/278/EEC

However, sewage sludges can also contain a wide range of organic micropollutants but PCBs and PAHs are considered to constitute the greatest health hazard of those normally encountered in sludges (see sections 6.3 and 6.4).

7.5 Hazardous wastes

7.5.1 The nature and amounts of hazardous waste produced

Hazardous wastes consist of individual waste materials and combinations of wastes that are presently, or potentially, hazardous to humans and other living organisms by means of their physical or chemical characteristics, the process by which they are produced, or their effect on human health or the environment (Manahan, 1984; BMA, 1991). The criteria used in the classification of hazardous substances include the type of hazard involved (toxicity, explosiveness, flammability and corrosiveness), the generic category of the substances (e.g. pesticides, wood preservatives, solvents, medicines), technological origins (electroplating, oil refining) and the presence of specific substances (e.g. Cd and Pb compounds, PCBs). Radioactive waste substances are normally classified separately owing to their high toxicity and long life. A brief list of some of the types of substances classified as hazardous includes explosives, compressed gases, flammable liquids, flammable solids, oxidizing materials, corrosive materials, poisonous materials, etiologic agents (e.g. bacteria and viruses causing infecious diseases), and radioactive materials (Manahan, 1991).

Hazardous wastes can cause considerable pollution of air, water and soil even before they are officially disposed of. This can occur during transit, such as when ships carrying hazardous wastes have sunk or have lost deck cargoes of dangerous chemicals overboard, and accidents involving lorries or trains carrying hazardous wastes. Widespread pollution has also occurred during the temporary storage of wastes in ponds and open tanks where they can volatilize in the air, or leak into surface water, or the soil and groundwater. However, in many cases these instances of pollution have not been covered by disposal legislation because that only applies when they are formally disposed of. New legislation in some countries now places an obligation on the generator of the waste to be responsible for it from its generation to its final legitimate disposal.

World total production of hazardous and special wastes = 338×10^6 t
 (of which 275×10^6 t, or 81%, generated in the USA)
Total production of hazardous and special wastes in OECD countries =
 303×10^6 t
(OECD, 1991).

Of the hazardous waste produced in the USA, 79% came from the chemical industries, petroleum refining produced 7%, the metal industries 2%, and 12% came from miscellaneous other sources.

7.5.2 Hazardous waste management

Source reduction of hazardous wastes Where possible, the industrial process producing the waste should be modified so that as far as possible the hazardous by-products are either avoided, re-used, or at least minimized. The latter approach is referred to as source reduction and has been shown to be economically worthwhile by the 3M Corporation (Minnesota, Mining and Manufacturing Corp.) in the USA. The company launched a 'pollution prevention pays' programme in 1975, which had saved the company $192 million within 10 years and reduced its effluent discharge by 3.7 billion litres, eliminated 10 000 t of water pollutants, 140 000 t of sludge and 90 000 t of air pollutants. However, source reduction is not being adopted as quickly as was expected, and the chemical and allied products industry, which accounts for roughly half of the hazardous waste generated in the USA, has not yet adopted this approach on a large scale. One of the main obstacles to this is the reluctance of competing companies to share information (World Resources Institute, 1992).

The procedures used to treat the hazardous wastes that are produced include:

Physical methods (Manahan, 1991)
 • phase separation (filtration/sedimentation)
 • phase transition (distillation, evaporation, physical precipitation)
 • phase transfer (extraction adsorption)
 • membrane separations (reverse osmosis, hyper and ultrafiltration)

Chemical methods (Manahan, 1991)
 • acid/base neutralization
 • chemical extraction and leaching

- chemical precipitation
- oxidation/reduction
- ion exchange
- electrolysis
- hydrolysis
- photolysis

Thermal treatment (see section 7.4.3)
- incineration
- wet oxidation
- plasma arc
- molten salt
- superheated water

Biotechnological methods (see section 7.4.3 and chapter 2)
Ocean dumping
Perpetual storage
- landfill
- underground injection
- salt formations
- arid region, unsaturated zone

Ocean dumping basically relies on dilution to minimize the hazard from waste substances but in some cases localized effects occur due to poor mixing as a result of stratifiation, winds, tides and currents. The worst affected marine ecosytems are those of estuaries and land-locked seas with restricted circulation, with polluting industries either discharging into rivers that drain into the seas or discharging directly into the sea. Examples of extensively polluted seas include the Irish Sea, the Baltic, parts of the North Sea, the Mediterranean, the Black Sea and others. The Irish Sea is the most radioactively polluted sea in the world due to discharges from the Sellafield nuclear reprocessing plant. This sea is also extensively polluted with organic micropollutants, including PCBs, and metals from industrial areas, such as Liverpool. In 1969, a large number of dead sea birds were found in the Irish Sea and it is thought that their deaths were caused, at least in part, by pollutants. Liquid hazardous wastes dumped in sealed containers offshore will leak out eventually and will be dispersed and diluted in due course. The extent to which pollutants accumulate in food chains in the open oceans is not fully understood although the ubiquitous occurrence of substances such as DDT in marine fauna indicates that this is likely to be significant. The London Dumping Convention of 1972 was intended to reduce the disposal of hazardous substances in the sea and it contained annexes which listed substances which should not be dumped at sea unless it could be demonstrated that they were only present in wastes in trace quantities or would rapidly be rendered harmless in the sea (Tolba *et al.*, 1992).

During 1986–87, 579 000 t of waste (from England and Wales) were dumped at sea. About half of this amount was hazardous waste and so marine dumping is the second most important means of disposal for hazardous waste, after landfilling. In comparison, during 1988, only 5500 t of hazardous waste were incinerated at sea (BMA, 1991). It is assumed that the plume of incinerator fumes reacted with the seawater and hence did not require elaborate clean-up before emission from the incinerator.

In England and Wales, in 1986–87, 83% of hazardous waste was disposed of by landfill, usually co-disposed with municipal wastes. Around 7% of these hazardous wastes were pretreated chemically or physically to neutralize or stabilize them and so reduce their hazardous properties before they were placed in landfills.

Only 2% of hazardous wastes produced in England and Wales were disposed of by incineration on land (BMA, 1991). However, some of the most highly toxic wastes, including those containing chlorinated hydrocarbons, were rendered harmless by this route. As discussed in section 6.4, the products of incomplete combustion of various chlorinated organic waste could include TCDD (dioxins) and PCDFs (furans).

7.5.3 New technologies for waste disposal (from World Resources Institute, 1992)

Biotechnology. This involves using appropriate microorganisms to decompose organic wastes which are too hazardous to treat by other means. This technique is more appropriate for the treatment of contaminated land, or stripped contaminated topsoil, than hazardous wastes from industrial processes. Although research has been conducted into the use of genetically engineered organisms for this, most of the working techniques involve the use of genetically adapted organisms isolated from sites which have been polluted for a relatively long time. The advantages of this method are that it is relatively inexpensive and does not require expensive specialized equipment. However, it is a relatively slow process which may not completely destroy all the hazardous organic compounds.

Plasma arc destruction. This technique creates temperatures of up to 45 000°C and a plasma state. It is particularly useful for destroying inert organic pollutants, such as PCBs and materials containing them. The main advantage of the method is the almost complete destruction of refractory hazardous organic substances but it is likely to be expensive.

Molten salt destruction. Hazardous chemicals, such as DDT, chemical warfare agents, corrosive solvents and acids can be destroyed by treatment in a hot bath of molten salt at 1650°C. Other methods involve a molten mixture of Na_2CO_3 and Na_2SO_4 at 900°C which has brought about the 99.999% destruction of hexachlorobenzene. This method does not require so much energy as some forms of incineration but is only around 99% effective over a range of compounds.

Superheated water. Water heated to 370°C at 220 bars pressure can dissolve normally insoluble organic chemicals and if oxygen is added to this water the organic pollutants are oxidized to CO_2 and H_2O. Inorganic compounds in the water combine to form salts. This method competes favourably with incineration for the destruction of most organic pollutants.

**7.6
Long-term pollution
problems of abandoned
landfills containing
hazardous wastes**

Many countries have abandoned landfills or areas of derelict land where hazardous wastes have either caused toxicity or other problems already, or where the risk of potential toxicity problems is so great that it cannot be tolerated and the site will need cleaning-up urgently.

7.6.1 Love Canal, New York, USA

The incident which drew attention to the problem and established hazardous wastes as a political problem was the Love Canal waste tip near Niagara Falls, in the State of New York, USA. This site, a section of an abandoned excavation for a canal, had been used from 1930 to 1952 for the disposal of wastes from a nearby chemical works. It had received about 20 000 tonnes of waste, comprising at least 248 different chemicals (including intermediates from the manufacture of chlorinated organic pesticides). Following its closure, the land was taken over in 1953 for building a school and a housing estate. Although various complaints had been made about irritation of children's eyes and respiratory tract, heavy rain and snow melt in the winter of 1977–78 caused groundwater containing chemicals to rise into the basements of houses built near to the tip and some buried drums to break through the covering soil. More severe health effects were reported and the US Environmental Protection Agency (EPA) carried out a series of investigations. Analysis of the groundwater revealed 82 different chemicals, 27 of which were on the EPA's Priority Pollutant List and 11 of these were known carcinogens. Several hazardous VOCs including benzene, toluene, chloroform, trichloroethylene, tetrachloroethylene, hexane and xylene were detected in the houses around the site. The site was declared a disaster area and 239 families were evacuated immediately followed by others later. This problem initiated surveys of other sites with serious pollution problems and it is interesting to note that Love Canal came only 25th in order of hazard. The EPA has documented at least 2100 sites used for the disposal of industrial wastes. The Love Canal incident prompted the CERCLA (Comprehensive Environmental Response, Compensation and Liability Act) of 1980, the 'Superfund' legislation. This is intended for the identification of severely polluted sites, their evaluation, monitoring and clean-up. The Love Canal site has cost many millions of dollars so far and has not yet been cleaned up (Manahan, 1984 and 1991; BMA, 1991) (see Table 6.19).

7.6.2   Lekkerkirk, near Rotterdam, The Netherlands

A similar problem to Love Canal was discovered in a new village called Lekkerkirk, near Rotterdam in the Netherlands. Many of the houses in this development were built adjacent to a river in 1970–71 on land raised by a layer of < 3.5 m of household and demolition waste covered with 0.7 m of sand. By 1980 it was realized that rising groundwater had carried pollutants upwards from the underlying wastes into the foundations of the houses. This caused deterioration of plastic drinking water pipes, contamination of the water, noxious odours inside the houses and toxicity symptoms in the garden plants. As a result of investigations, 250 houses had to be abandoned while 87 000 m^3 of contaminated fill containing 1600 drums of chemicals comprising mainly paint solvents and resins containing toluene, lower boiling point solvents and Cd, Sb, Hg, Pb and Zn were removed and transported by barges to Rotterdam for destruction by incineration. The cost of cleaning the site up in 1981 was the equivalent of £156 million (Manahan, 1984; Finnecy and Pearce, 1986; BMA, 1991).

An OECD environmental study (1991) reported the following estimated numbers of abandoned waste tips in Europe, although not all will have received hazardous wastes (OECD, 1991):

Austria	– around 1000 sites;
Netherlands	– 5000 sites identified with 350 requiring immediate remediation;
former West Germany	– 21 000 abandoned landfills with 2000 requiring clean-up;
Denmark	– 380 sites being cleaned up in 1987;
France	– 66 abandoned sites ('points noirs') posing grave risks to human health.

7.7
Tanker accidents and oil spillages at sea

Although not strictly waste disposal, oil spillages at sea are important pollution events and the oil slicks produced are quite spectacular and so attract the attention of the world news media. However, it has been estimated that less than 4% of the total amount of oil discharged into the sea is the result of tanker accidents, while 22.5% is from operational

Figure 7.1   Abandoned Cu mine at Parys Mountain in Anglesey, Wales (photo: B.J. Alloway).

discharges. The 1982 estimate for oil entering the sea was 3.1×10^6 t of which up to 20% was from natural sources outside human control, 45% enters from land via rivers and from coastal refineries, etc. Transportation was responsible for up to 40%, with oil tankers accounting for 20–25% of total (Gourlay, 1988). Nevertheless, spillages associated with tanker accidents do cause a lot of ecological and economic damage, due to the concentrated effect of the slick, which is often near to a shoreline which affects both commercial fisheries and tourism as well as the marine ecosystem.

The largest spills to date have been from the *Atlantic Express* which released 276 000 t and affected Tobago in 1979, *Castello Belver* (256 000 t) which affected South Africa in 1983, and *Amoco Cadiz* (223 000 t) which affected Brittany, France, in 1978 (Tolba *et al.*, 1992). In comparison with these the *Exxon Valdeez* spillage of 45 000 t in Alaska and most recently the *Brear* (85 000 t) off the Shetland Isles, Scotland, appear much smaller although their local ecological and economic effects were considerable. It is interesting to note that the number of oil-spills greater than 6800 t actually decreased by 74% between 1974 and 1986 (Tolba *et al.*, 1992). The intentional discharge of oil into the Persian Gulf from Kuwait by Iraqi occupying forces also caused similar ecological damage. These same forces also set fire to many oil wells and these took quite a long time to put out and prevent the loss of oil. The air pollution, mainly smoke from the oil well fires, caused the pollution of a large area of land with fall-out from the smoke (containing PAHs, etc.) and also affected weather patterns.

The spread of oil slicks produced by spillages into the sea depends on the type of oil, how dense it is, how much there is, the tidal, wave and wind conditions, and the rate at which the volatile components evaporate and the denser residue breaks up to form floating lumps. High temperatures and rough seas promote this process and the slick from the *Brear* in the Shetlands was certainly strongly affected by the gale force winds which persisted for several days after the accident. Some aromatic hydrocarbons dissolve in the water under a slick and increase the ecotoxic effects of the slick. A lot of the oil forms an emulsion with seawater known as a 'mousse' which eventually washes up on beaches as tar-balls. The hydrocarbons are subject to oxidation catalysed by UV light, and microbial decomposition will also occur. The ecosystems of the sea floor and the inter-tidal zone will all be affected by the oil, although it is often claimed that the dispersants which have been used to break up the slicks have sometimes caused a lot of ecotoxicity also. Sea bird and mammal populations are also badly affected. However, some studies on ecosystems affected by major oil spills have indicated that recovery has been more rapid than was expected.

As stated above, the majority of the oil entering the sea from tankers is from operational discharges, such as when the tanks and pipe are flushed out with sea water before loading with fresh crude oil. This is more dispersed than the slicks from spillages and has less local impact although floating oil mousse does get washed up on beaches all over the world especially those near main shipping routes.

7.8
Other multi-pollutant situations

For the world as a whole, it is possible to generalize and say that the majority of pollution situations involve more than one pollutant although in some cases one or a few compounds may predominate (see chapter 2). The multiple pollutant situation has several implications: in some cases some of the pollutants may have additive or even synergistic effects and in others they may tend to counteract each other's bioaccumulation or harmful effects. Waste disposal is an obvious source of a wide range of pollutants, but there are many other multi-pollutant sources. Derelict land, defined in the UK by the Department of the Environment as 'land so damaged by industrial or other development that it is incapable of beneficial use without treatment', is in almost all cases a multiple pollutant problem. For example, an old gas works site will be polluted with tars (containing PAHs, benzene, xylene and napthalene), HCN, phenols, As, Pb, Cu, cyanides, sulphates and sulphides (see section 6.4). Scrapyards are polluted with a range of metals, PCBs, PAHs, hydrocarbons and the products of the partial combustion of plastics.

Even a normally occurring natural hazard such as flooding is a pollution event. Sewers overflow, drums and tanks of various substances in works, farms or even houses leak or burst, and so the water carries a mixture of pollutants (although often dilute) which are deposited on the land that the floodwater covers.

Land used for military training, munition and equipment stores and the battlefields themselves are either sources or sinks of a wide range of pollutants including hydrocarbons from fuel spillages, explosives, mines, bullets, smoke generating compounds, de-icer, and, possibly, chemical warfare agents (including nerve gases,) and herbicides (e.g. 'Agent Orange', see section 6.4) and many others including PAHs and other products of incomplete combustion. Many former military bases and training grounds have been abandoned in central and eastern Europe and need careful cleaning up. Recent wars and skirmishes around the world will have left a large amount of modern warfare materials to pollute the environment, crashed aircraft will have caused localized pollution as does even a stolen car which is burnt in a field in any 'peaceful' country. The battlefields of the First and Second World Wars will still be polluted with persistent pollutants, especially heavy metals which have half-lives of hundreds, even thousands of years.

7.9
Chemical time bombs

The concept of chemical time bombs proposed by Stigliani (1991) and others is a useful way of focusing attention on situations where hazardous chemicals may suddenly be released into the environment. As listed by Tolba *et al.* (1992) these can include rapid acidification of soils and lakes due to depletion of the buffering capacities of forest soils, leaching of phosphates from agricultural soils into aquatic ecosystems resulting from saturation of sorptive mechanisms, leaching and plant uptake of metals from polluted agricultural soils due to acidification or reduced sorption capacities, metals in coastal waters and estuaries due to desorption from sediments, and the release of sulphuric acid and metals from wetlands or drying out due to drainage or climatic change. In addition to these general types of sources, individual sources of pollutants, such as hazardous wastes in containers dumped in the sea or in landfills may also cause a time bomb-like release of pollutants as exemplified by Love Canal (section 7.6.1) and Minamata (section 5.3).

References

British Medical Association (1991) *Hazardous Waste and Human Health*, Oxford University Press, Oxford.

Council of the European Communities (1986) Directive 86/278/EEC on the Protection of the Environment and in Particular of the Soil, when sewage sludge is used. EEC, Brussels.

Ellis, D. (1989) *Environments at Risk*, Springer-Verlag, Berlin.

Finnecy, E. E. and Pearce, K. W. (1986) in Hester, R. E. (ed) *Understanding Our Environment* (1st edition), Royal Society of Chemistry, London.

Fish, H. (1992) Chapter 3 in Harrison, R. M. (ed) *Understanding Our Environment* (2nd edition) Royal Society of Chemistry, Cambridge.

Gourlay, K. A. (1988) *Poisoners of the Seas*, Zed Books, London.

House of Commons Environmental Committee (1989) 1st Report, *Contaminated Land*, HMSO, London.

Lester, J. E. (1990) in Harrison, R. M. (ed) *Pollution: causes, effects and control* (2nd edition) Royal Society of Chemistry, Cambridge.

Ministry of Agriculture, Fisheries and Food and Welsh Office Agriculture Department (1992) *Code of Good Agricultural Practice for the Protection of Soil*, Draft Consultation Document, MAFF, London.

Manahan, S. E. (1984) *Environmental Chemistry* (4th edition) Brooks/Cole (Wadsworth), Monterey, Ca.

Manahan, S. E. (1991) *Environmental Chemistry* (5th edition) Lewis Publishers, Chelsea, Mich.

Neutra, R. R. and B. Ostro (1992) *Sci. Total Environ.*, **127** 91.

Organisation for Economic Co-operation and Development (OECD) (1991) *The State of the Environment*, OECD, Paris.

Stigliani, W. M. (ed) (1991) *Chemical Time Bombs: Definitions, Concepts and Examples*, IIASA, Laxenburg, Austria.

Tolba, M. K., El-Kholy Osama, A. El-Hinnawi, E., Holdgate, M. W., McMichael, D. F. and Munn, R. E. (eds) (1992) *The World Environment 1972–1992*, UNEP, Chapman and Hall, London.

World Resources Institute (1992) *World Resources 1992–93*, Oxford University Press, Oxford.

Appendix Table of units and conversions

SI prefixes

10	deca (da)	10^{-1}	deci (d)
10^2	hecto (h)	10^{-2}	centi (c)
10^3	kilo (k)	10^{-3}	milli (m)
10^6	mega (M)	10^{-6}	micro (μ)
10^9	giga (G)	10^{-9}	nano (n)
10^{12}	tera (T)	10^{-12}	pico (p)
10^{15}	peta (P)	10^{-15}	femto (f)

SI units

length	metre (m)
mass	kilogram (kg)
time	seconds (s)
temperature	kelvin (K)
force	newton (N)
pressure	pascal (P)
energy	joule (J)
power	watt (W)
heat	joule (J)

In the following table the larger the unit the smaller it appears in magnitude relative to the left hand column. Equivalence of a pair of units is given by their ratio, for example:

cm^3/litre $= 1/10^{-3} = 1000$ lb/kg $= 2.2 \times 10^{-3}/10^{-3} = 2.2$

Volume

cm^3	litre	m^3	ft^3
1	10^{-3}	10^{-6}	3.53×10^{-5}

Mass

g	kg	lb	t (metric)
1	10^{-3}	2.2×10^{-3}	10^{-6}

Energy

erg	joule	kW	Mev
1	10^{-7}	2.78×10^{-14}	6.24×10^{-5}

Pressure

Pa	dyne/cm^2	p.s.i.	atmospheres
1	10	1.45×10^{-4}	9.87×10^{-6}

Area

cm^2	ft^2	acre	hectare
1	1.08×10^{-3}	2.47×10^{-8}	10^{-8}

Index

abortion rate, Lombardy 242
accepted level of vinyl chloride
    in workplace 216
acetic acid
    solution pH 131–2
    USA production 210
acetone, USA production 210
acetylcholine 249
activity coefficients 64
adhesives 211–2
adiabatic lapse rate 25
advanced gas-cooled reactor
    (AGR) 175
aerosols
    formulation 213
    production 213
Aflatoxin 47
Agent Orange 237–9, 243
aggregated dead zone (ADZ)
    32
agriculture as pollution source
    18
air pollution
    analysis 81–4, 102
    from well fires 277
air sampling, direct 76–7
Aldrin 85
    cross resistance 228
    in drinking water 57
    in Lake Kariba 229
    synthesis 225
    toxicity to fish 226
algae in food chain 228–9
Alkali Act (1862) 6
alkali metals by AAS 93
alkanes 203–4
    indoor levels 221
    in urban air 205
    metabolism 215

Allethrin 246, 247
alpha particle 168, 185
Altzheimerr's disease 165
aluminium
    in rain 134
    in water 164
alveoli 151
Ames test 48, 49
ammonium nitrate 134
Amoco Cadiz 276
Amsterdam, transport 127
anaemia, lead induced 163
analytical quality assurance
    97–8
ancient civilizations 140
Antartica 116
    ice 122
    metal levels 144
    ozone depletion 111, 113
    pollution 21
    the seal 234
anthracene, structure 206
anthracite 200
aphids
    parathion against 250
    resistant 254
aquifers
    and groundwater 32
    leachate pollution 265
Arctic ocean
    brown snow 41
    pollution 21
    seal 234
arene oxides 210
Aroclor 232
aromatic compounds,
    metabolism 209, 210
aromatic hydrocarbons,
    indoors 221

arsenic
    LD_{50} in mammals 157
    principal features 159–60
    soil quality criteria 160
asbestos 188
    in Hg cell 222
    in USA 190
    types 189–90
asbestosis 190
asthma 191, 199
Atlantic Express 276
atomic absorption spectroscopy
    86–98
    basic theory 88–9
    calibration curve 90
    double beam 91
    instrumentation 91
    interferences 92–3, 97
    of lithium 88
Atrazine 22, 40, 41, 254
Australian seal, PCB in milk
    234
Austria, cancer incidence in
    livestock 148
aviation and ozone 112
Avogadro's number 173
Avonmouth smelter, Cd levels
    149

bacteria
    and pH 37
    and soil acidity 134
    Clostridium sp. 45
    decomposition by 40, 41
    Fe and Mn transforming 39
    for methylation 154
    in effluent 220
    Salmonella sp. 48
barium 169

batteries 163
  metal content 149
battlefields 278
Becquerel, definition 185
Bedfordshire, brickmaking
  Mo from 167
  odour from 255
benzene
  in Lindane synthesis 223,
    224
  production in UK 216
  production in USA 210
  toxicity 216
benzo-[a]-pyrene 81
  carcenogenicity 207
  in drinking water 57
  structure and activity 206
benzo-[e]-pyrene, structure
  and activity 206
benzthiazole, urban pollution
  by 77
benzyl radical 205
beryllium
  in coal 165
  mammalian LD_{50} 157
beta particle 168, 169
BHC, persistence 230
Bilthoven, air pollution 71, 72
Binghampton, New York
  234–5
biomagnification 228–9
bioresmethrin, structure and
  toxicity 246, 247
bitumen, use and run-off 208
blood, lead levels 164
BOD 6, 33–4
  of sewage 269
boiling water reactor (BWR)
  175
Bordeaux mixture 161
boric acid, as moderator 175
Brear 4, 276
breeder reactor 175, 177
brevitoxin, and death of
  dolphins 234
brick kilns, as source of F 166
British Coal Corporation 139
buffer action 132, 134
builders, for detergents 219

butadiene, USA production
  210

cabbage, rootfly, resistance to
  DDT 228–9
cacodylic acid, in humans 238
cadmium
  in diet 149
  mammalian LD_{50} 157
  principal features 160–3
  replacing Ca 39
^{137}caesium 177–8
calcium sulphate 138, 139
calcium, sequestration 220
calibration of GLC detectors
  66
Californian seal, PCB in milk
  234
Camelford, water supply 165
Canada
  Environmental Quality
    Criteria 54–6
  polluted soil classification
    158, 160
cancer 189, 216
  adrenal 184
  bladder 5, 269
  causation 47, 49–50
  from beryllium 166
  from Pu 178
  from vinyl chloride 48, 216,
    231
  mortality in UK 207
  of lung 162, 185, 198
  of lung and bowel 190
  of skin 112, 114, 206
  of thyroid 184
  respiratory 160
  scrotal 198
  tumour types 48
Cancer, International Agency
  for Research on 208
CANDU reactor 175–6
capilliary columns for GLC,
  advantages 70
carbamate pesticides 253–5
  mechanism of action 250
carbaryl 253–4
  from phosgene 253

carbazole, in smoke 209
carbon dioxide 116–124
  carbon labelling 136
  emissions by fuel type 118
  emissions/capita 118
  increase in levels 116–7, 121
  indoor level 221
  toxicity 57
  trapping of infrared 119
carbon monoxide
  indoor levels 221
  toxicity 57
carbon tetrachloride 214, 222
  cancer from 217
  for cleaning 214
  metabolism 217
  production 215
carbonyl compounds, in air
  203, 205
carcinogenecity of dioxins 239
carcinogens 23, 24
Castello Belver 276
catalytic convertion 128, 208
cation exchange capacity
  (CEC) 37
CEGB 132, 138
cellulose 199
cement, sodium in 93
cerebral oedema 164
CFC 12 74, 76
chalcopyrite 146
chemical ionisation for GC/MS
  76
chemical time bombs 3, 278
chemical warfare agents 278
chemicals, mammalian
  exposure 44
Chernobyl 178, 183–4
chimney plumes 26–29
China
  population pressure 118, 119
  soil transport 41
Chlorpyrifos 251
  toxicity 253
chlor-alkali process 162, 221
Chloral 224
chlorinated hydrocarbons, in
  soil 53
chlorinated solvents, in soil 40

chlorine 221–3, 269
chlorobenzenes 233
chlorobiphenylenes 235, 236
chlorofluorocarbons 76, 112,
 114
  control 115
  for cleaning 214
  infrared spectrum 119
  manufacture 112
  nomenclature 113
  photolysis 112
  stratospheric 112
chloroform
  in drinking water 57, 216
  in environment 216
  reaction in incinerators 243
  WHO guideline 216
chloronaphthalenes 233
cholera 222, 267, 268
Choline 249
cholinesterase 253
  mechanism of action 249–50
Chromatography 59   see also
  TLC, GLC, HPLC
  column 59
chromium
  in coal ash 148
  principal features 162
chrysanthemic acid 245
Chrysanthemum
  cinerariaefolium 244
chrysene
  in air 81
  structure and activity 206
cigarette smoke 221
cinerolone 245
clean air 116
Clean Air Act (1956) 6, 197,
  198
Clear Lake, California 228
Clophen 232
Clostridium botulinum 45
coal
  consumption worldwide 130
  composition 200
  dust 191
  fluorine content 167
  formation 199
  production 202

sulphur content 129
  tar industry 207
coatings 212
^{60}cobalt 170
Coke and Coalite Company
  237
cold fusion 188
comparison of GLC and
  HPLC 83
compost, metal content 147,
  150
composting 266
contaminating uses of land 17
control zones at Séveso 240–1
Copenhagen, International
  Meeting 114
copper
  ores 145–6
  principal features 161
Corbicula africana 228–9
cormorant 229
crocodile 229
cryolite 166
Cunninghamella elegans, attack
  of aromatics 41
Curie, M. and J. 167
cyanide on derelict sites 54
cyclohexane, USA production
  210
Czechoslovakia, pollution 9
Czechoslovakian coal 129

Dalapon 226, 227
Darcy's Law for flow of
  groundwater 32
DDD
  in food chain 228
  structure 224
DDT
  human toxicity 227, 238
  in drinking water 57
  in Lake Kariba 229
  in Third World 10
  isomers, synthesis 222–4
  persistence 230
  resistant strains 227
  trace enrichment 85
deltamethrin 247
denitrification, of sewage 268

dental alloys 149
derelict land, definition 277
derivatization 80, 81, 88, 86
Dermatophagoides
  pteronyssinus, in dust 191
Derris elliptica 248
desertification 123
desorption, of air samples 71
detection limits, for metals 98
detergents 217–220
  synthesis 218
  types 219
  usage in USA 218
dibenzofurans 235–6
dichlorophenoxyacetic acid
  (2,4–D) 226, 227, 238
Dichlorvos 251
  toxicity 253
Dieldrin
  in drinking water 57
  persistence 230
  synthesis 225
  toxicity 226
  trace enrichment 85
Diels–Alder reaction 225
diesel emission 215
diffusion, during
  chromatography 69
dinitro-o-cresol 226, 227
dioxin detection limit 240
dioxins 235–242
direct air sampling 76–7
  pollution contours 77
DNA 207
  structure 49
DOE (UK) 179
  Committee for Reclamation
    of Contaminated Land
    158–9
  red list 21, 22, 254
  trigger concentrations 53–4
Doll, Sir Richard 207, 239
dolphin deaths 234
dose response, to toxicants 44
Drax power station 138
drinking water   see also water
  carbon tetrachloride in 217
  chloroform in 216
  disinfection 269

herbicides in 42
metals in 152
pH of 153
pollutant guidelines 55
re-use 268
risk from pipes 188, 190, 275
zinc in 164–5
dry cleaning, of clothes 214–5

earth's crust, composition 142
EC
  adhesives production 211
  Common Agriculture Policy
    33
  control of heavy metals 270
  directive for SO_2 and NO_x
    137, 140
  directive on cadmium 149,
    161
  directive on smoke 198
  forest damage 135
  levels of metals 144
  limits for metals in soil 270
  limits for pesticides 253
  population 10
  standards for car emissions
    128
ecotoxicology 50–1
effective stack height 28
electric cars 129
electron capture detector 66
electron volt 88–9, 169
electronics, metal components
  149, 150
emphysema 165
enantiomers, of pyrethrins 245
endosulphan 225, 226
enerrgy
  critical 171–2
  of emission 170
  use and GNP 11
  yield from fission 172–3
environmental monitoring
  99–104
Environmental Protection
    Agency (USA) 239, 242
    see also EPA and USA
enzymes

action of acetyl choline
    249–50
  mechanism of action 251
  oxygenase 41
EPA
  hazardous substances of
    classification 17, 19, 20
  hazardous waste 271
  priority pollutants 274
  Superfund 275
epoxidation
  of Aldrin 225
  of PAH 207
  of vinyl chloride 232
ethanol, in humans 238
ethylene glycol, USA
    production 210
ethylene, USA production 210
European Community see EC
excited state
  of lithium 88
  temperature variation 89
Exxon Valdeez 276

ferrihydrite, sorption by 39
fertilizers, metal content 147
fish, chromium poisoning 162
fish, LC_{50} of
    organophosphorus
    compounds 253
fish, pyrethrin LD_{50} 247
flame ionization detector 65
fluidized bed combustion 139
fluoranthrene, structure and
    activity 206
fluorine
  in tea plant 165, 167
  industrial uses 166
  in water supplies 167
fluorosis 167
Fluorspar 147, 166
food chains
  in Clear Lake and Lake
    Kariba 228
  mercury in 154
  strontium 90 inclusion in 178
forest damage 9
  by NO_x 126
  in EC 135

in West Germany 134
Forestry Commission 134
formaldehyde
  formation 203
  indoor levels 221
fossil fuel
  additives 149
  emissions from combustion
    197
  metal content 147–8
  usage 121
fractionation of petroleum
  201–2
France, SO_2 and NO_x
    production 137
free radicals
  action on PAH 207
  and PVC production 231,
    232
  initiation of PAN 204
Freundlich isotherm 154
fungicides, metal content 147

galena 164
gamma radiation 169, 170
gangue minerals 147
gas works
  abandoned 277
  as pollution sources 18
gases, dangerous 57
Gaussian curve 67
Gaussian model for pollution
    transport 27–9, 31
Geneva Climate Conference
    122
Germany, SO_2 and NO_x
    production 137
GLC 62–77
  column packing 63
  comparison with HPLC 83
  essential hardware 62
  of chlorophenols 86–7
  of urban air pollutants 71–2,
    76–7
  principle parameters 67
  progress of separation 64
  relative retention 65
  temperature programming
    68–9

glue sniffing   *see* solvent abuse
glutathione 210
glyphosate 252, 253
GNP and energy use 11
grebes, death 228
greenhouse effect 119–123
groundwater, pollution by
   organics 41, 42

half-life
  of atrazine 254
  of pyrethrins 247
  of radionuclides 168–70, 185
  of sulphur 35, 171
hardness of water 218, 219
Harrisburg nuclear accident
   181
hazardous waste 263, 270–5
  classification 270
  management 271
heart disease 248
heavy metals 140–164
  adsorption on soils 38
  analysis 158–64
  and fish 133
  biochemical properties 141–2
  coprecipitation 38–9
  defined 140
  EC limits 270
  in air and water 144
  in rocks 143
  in sludge 269–70
  in soils 103, 153–4
  methylation 154
  mobilization 220
  primary production 141
  quality criteria 55
  sources 142–3, 145–9
  toxic effects 141, 150, 155–8
  uptake by plants 154, 156
  urban pollution by 275
Heptachlor 225
H.M. Inspectorate of Pollution
   242
herbicides 226–7
  in fish 237
  toxic components 235–42
herring gull 234
HETP 67

hexachlorobenzene 243
  in drinking water 57
n-hexane, persistence and
  toxicity 216
Hofmann La Roche 241
homeostasis 142, 155
house dust mite 191
house fly, resistance to DDT
  227
House of Commons
  Environment Committee
  17
HPLC 77–86
  comparison with GLC 83
  components 78
  detectors 80
  of amines 83–4
  of phenols 81, 83
  reverse phase 78–9
  sample injection 78, 80
  solvents 79
  of PAH 82
human diet 155, 157
human food chain 160, 162
human sperm 5
humic polymers 36, 37
hydrocarbons in smog 72   *see*
  also alkanes
hydrofluorochlorocarbons for
  cleaning 214
hydrogen fluoride, discharge to
  air 166, 198
hydrogen sulphide
  in petroleum 200
  toxicity 57
hydroxyl radical 126, 130, 235,
  236
  reaction with alkanes 202–4
  reaction with aromatic
   hydrocarbons 205

infrared absorption spectra
  of carbon dioxide 120
  of methane 119, 202
  of nitrous oxide 119
  of water vapour 120
ice cores 117
ICI 254
ICMESA plant 237

igneous rocks 143
incineration 235, 265–6
  at sea 273
  of organochlorines 242–4
India, energy demand 118, 119
indolylacetic acid 226, 227
indoor pollution 220–1
  by radon 169
  by smoke 198
  models 30–1
  sources 23
inks 212
inorganic pollutants, in soil 53
insecticidal strips 251
insecticides, toxicity 227–9
internal standard, for GLC 86
International Atomic Energy
  Agency (Vienna) 179
inversion, atmospheric 25–6,
  29
iodine, radioactive 177, 180,
  183
iodophenphos 251
  toxicity 253
ion exchange 37–8
ionization potential 90
Iraq, human poisoning 163
Irish Sea pollution 272
irritation, by fibres 188
isocratic elution 84
isomers of PCBs 232–3
itai-itai disease 161
Italy, SO_2 and NO_x production
  137

Japan
  adhesive production 211
  human poisoning 162
  metal pollution 161
  waste production 263, 265
Jintsu Valley pollution 161
Joint European Torus (JET)
  187

Kapenta 228–9
Kennaway, Sir Ernest 207
kerogen 200
kidney damage
  by lead 163

by organochlorines 216, 217
Kiev 183
Kuwait, pollution 276

Labeo attivelis 229
Lake Constance, phosphate
    pollution 220
Lake Kariba, DDT levels
    228–9
Lambert–Beer Law 90
land contamination 17
landfills 264–5
    abandoned 274–5
    cost 243
Langmuir isotherm 153
lead
    distribution 5
    in particles 148
    mammalian LD_{50} 157
    on foliage 151
    principle features 163–4
    waste 214
leaf damage, by PAN 204
Lekkerkirk, polluted site 275
leukemia, myeloid 216
lichen deserts 134
lignin, structure 199
    in waste 243
lignite 9
    composition 200
limit value 52
Limits to Growth 8
Lindane 223–4
    cross resistance 228
    in drinking water 57
London
    odour nuisance 266
    transport 127
    smog 197–8
Los Angeles
    pollution 26
    vehicle usage 126–7
Love Canal, USA 237, 274–5
lung disease 165
Luxembourg, forest damage
    135
lymphomas 216

magnesium, sequestration 220

make-up gas 67
malaria 230–1
Malathion 115
    toxicity 253
    uses 251
Mars 120
mass spectra 73–4
    accurate mass 74–6
    and ICP 97
    as detector 71
    detection limits 98
    link to GLC 75
    of organochlorines 73–4
mass transfer 69
Mauna Loa 116, 117
maximum acceptable risk 179
medicines, metals from 149
melanoma 112
Melbourne, transport 127
mercury
    by electrolysis 221–2
    in paper pulp mills 162
    in seed dressing 162
    mammalian LD_{50} 157
    principal features 162–3
metal catalysts 149
metal cleaning, by solvents 214
metal ions
    cation exchange 153
    coprecipitation 151–2
    half-lives 153
    in drinking water 152
Metallogenum spp. 39
metallothionen proteins 142,
    155
metallurgical industry 18
metals   *see under individual*
      metals
    from wastes 149, 150
    heavy 140–164
    in scrap 148
    in sea water 144
    non-ferous, sources 146
    toxicity to mammals 157
methane 202–3
    from waste 265, 266
    infrared absorption 119
    toxicity 57
methanol, USA production 210

methoxychlor
    by trace enrichment 85
    structure 224
methyl bromide in stratosphere
    115
methyl radical 202–4
methylation of metals 37, 154
methyldichlorophenoxyacetic
    acid (MCPA) 226, 227
Metropolitan Edison company
    181
Meuse Valley, pollution 131
mice
    cancer from $CHCl_3$ 216
    effect of PAH on 207
    effect of TCDD 239
microbial decompositoin 277
micronutrients 161–4
    dose response 142
    essential metals 141
microorganisms 13, 244, 267
    and organochlorines 40–1
    effect of metals 155
    in soils 35, 36
    in water 34
    methylating 37
    waste disposal by 273
Milan 241
Minamata Bay, mercury
    mobilization 162
mine waste 263
mineral fibres 188–91
    analysis 189
mining, metalliferous 145
Minnesota Mining and
    Manufacturing
    Corporation 271
mixed function oxidase (MFO)
    251, 252
molecular extinction coefficient
    80
molecular ion 72
molten salt, waste disposal by
    273
molybdenum in livestock 167
MONSANTO Company 231,
    237
Montreal Protocol (1991),
    solvents 211, 214

Moscow, transport 127
motorway, benzene level 216
multi-pollutant situations 277–278
multiplication factor, for neutrons 172, 174
Munich, transport 127
municipal incinerators 243, 244
muon, the 188
mutagenesis 48
mycotoxins 23

naphthalene, structure 206
NASA 120
National Resources Defence Council (USA) 213
natural enrichment of metals 143
natural gas 201, 202
neptunium 171
nerve poisons 243, 248
Netherlands
    forest damage 135
    polluted sites 275
    quality standards 51–3
    toxicity evaluation 155–8
neurotoxicity 162, 163, 215, 234
neutron bombardment 171
    fast 175
    fission 172–3
New York
    transport 127
    risk from radon 185
nicotine 198
    toxicity 248, 249
NISSAN company 214
nitrate pollution, of soil and water 33
nitric oxide 125
    reactions 130
nitriloacetic acid (NTA) 220
nitrogen compounds in petroleum 201
nitrogen fixation 114, 125, 139
nitrogen oxides   see also oxides of nitrogen

control 137
indoor levels 221
production in EC 137
reactions in air 204
nitrogen peroxide 125
nitropyrene 208
nitrosoamines 33
nitrous oxide
    infrared spectrum 119
    lifetime 126
    production 124
North Pole 122
Norway, metal deposition 148, 151
nuclear accidents
    Chernobyl 183–4
    in Urals 180–1
    scale of risk 179
    Three Mile Island 181–3
    Windscale (Sellafield) 180
nuclear fission 171–8   see also nuclear power
    criteria 171–2
    ideal 172
nuclear power   see also nuclear reactors, nuclear fission
    from fusion 186–8
    in developed nations 186
    in the future 178–9
    social aspects 186
nuclear reactors 173–7   see also nuclear power
    fission products 177–8
    fuel preparation 173
    in UK 181
    in USA 180

ocean dumping 272–3
odd oxygen 110
odours 255–7
    control 256
    indoors 275
    perception 211
    threshold, of n-hexane 215
OECD countries
    abandoned waste 275
    hazardous waste 271
    waste production 263–6

oil
    consumption 130
    slicks 4
    spillage at sea 275–7
Ontario, PCDD emissions 243
organic pollutants
    degradation in soil 39–42
    quality criteria 55
organic solvents 210–217
    applications 211
    production 210
organochlorines 216–7
    and soil organisms 40
    disposal 242–4, 273, 274
    in groundwater 41
    mass spectra 73–4
organohalides 221–32
organophosphorus
    nerve agents, incineration 243
    pesticides 248–53
otter, decline of 234
Oxford University, building damage 137
oxidation, microbial 41
oxides of nitrogen 124–9   see also nitrogen oxides
    and acid rain 126–7
    and ozone depletion 126
    control 128, 139
    levels in air 125
    sources 124
ozone 109–15
    and SO_2 130
    as disinfectant 269
    control 115
    depletion 213
    depletion potential 113
    diurnal variation 115, 205
    formation 110–1
    indoor level 221
    infrared spectrum 119
    leaf damage by 136
    physical properties 110
    stratospheric 111
    toxicity 115
    tropospheric 114–5, 204, 212
    ultraviolet absorption 110–1

ozone layer 110
  depletion 111
  effect of CFCs 112

P-450 cytochrome 210, 242
paddy rice 161
PAH 206–9
  analysis 81, 82
  in particles 208
  in smoke 12, 198
  in soil 53, 54
  production worldwide 208
  sources 208
  toxicity 207
paints, metal content 149, 150
PAN
  in smog 203
  in urban air 205
  natural occurrence 204
paraquat 36
Parathion
  synthesis 149–50
  toxicity 253
Paris, transport 127
particles 188–91
  diameter 189, 196
  in atmosphere 151
  ingestion 189
  lead containing 163
  of beryllium 166
  sources 190
  toxicity 190
PCBs 232–8
  in air worldwide 235
  in seal milk 234
  occurrence 233
  pollution incidents 234
  synthesis 232–3
PCDD
  in humans 235
  incineration 243–4, 266
PCDF
  from PCBs 236
  in humans 235
  incineration 266
peat, composition 200
pentachlorophenol
  in drinking water 57
  in lake water 86, 87

peptides 249–50
peregrine falcon, UK
  population 230
permethrin 246, 247
peroxyacetyl nitrate   see PAN
perylene, structure and activity
  206
pesticides 221–32, 244–55
  control of 229–31
  in current use 21, 22
  in soil 53
  metal content 147
  natural 244–48
  persistence 230
petrol
  critical concentration 57
  emission 215
petroleum 200–2
  consumption 200
  formation 200
  properties 200–1
  release to environment 202
  reserves 7
Phanaerochaete chrysosporium
  and effluent 244
phenanthrene, structure 206
Phenoclor 232
phenols
  analysis 81–3
  as metabolites 210
  chloro- 86–7
  formation 205
  USA production 210
Phosgene 217, 253
phosphate, in detergents
  219–20
photolysis
  and air pollution 204–5
  of NO_2 114
  of ozone 110
  of pyrethrins 246
  on soil 36
pH, typical values 131–2
Physostigmine 253, 254
phytochelatins 141, 154
phytotoxicity
  of Cr(VI) 162
  of metals 155–6
pica 151

picloram, in humans 238
Pittsburg, SO_2 pollution 131
Planck's equation 89
plasma arc, for waste disposal
  273
plastics
  hot melting 212
  metal content 149
plumes
  pollution contours 77
  tracing 171
  transport and dispersion
  26–9
plutonium 171
  fission products 177–78
  mammalian LD_{50} 157
  toxicity 155
pneumoconiosis 166, 189, 191
Point Barrow 116
pollutants   see also pollution
  effect on man 47
  in contaminated land 51
  model pathway 16
  reactions in air 31
  sources 17–9, 29–30
  subclinical effects on animals
  47
  sub-lethal effects 50–1
  transport 24–34
  types 6, 7
pollution, definition 5
  of rivers 32
  types 12, 13
polonium 169, 185
polybromobiphenyl 233
polychloroterphenyls 233
polycyclic aromatic
  hydrocarbons   see PAH
polymerization of vinyl
  chloride 231–2
polystyrene in waste 243
polyvinyl chloride (PVC)
  231–2
  in waste 243
population
  and energy use 118
  growth 9, 10
  levels 118
Portland cement, analysis 93

potassium, radioactive 170
power station 130, 138
   as pollution sources 18
   emissions of NO_x 129
pre-column for HPLC 85
predatory birds 228–9
   peregrine falcon 230
pressurized water reactor
      (PWR) 174–5
principle parameters for GLC
      67–70
printing, metals from 149
production
   of PVC 231–2
   of pyrethrins 244
   of vinyl chloride 231
proinsecticides 251
propylene, USA production
      210
protactinium 168
pyrene, structure and toxicity 206
pyrethric acid 245
pyrethrins 244–6
   criteria for activity 246
   synthetic 246–48
pyrethrolone 245
pyrimicarb 254
pyrite from sulphate 39
pyrite, oxidation 145

quality goals, Netherlands soils
      157–8
quantum theory 189
Quebec, incineration study 243

radioactivity
   extreme exposure 183
   in buildings 184–5
   protection from 169–70, 185
   transport 184
rad, the 169
rats
   and LD_{50} of
      organophosphorus
      compounds 253
   and LD_{50} of pyrethrin 247
   cancer and effect of $CHCl_3$
      216
   effect of PAH on 207

effect of TCDD on 239
red list (UK) 21, 22, 254
red spider mite 251
rem, the 170, 183
resistant strains 227
resmethrin 246
response factor in GLC 86
Rhizobium bacteria 124
rice bran contamination 233
risk
   assessment of toxic 51–6
   evaluation 156–8
   from radon 185
Rivers Pollution Act (1876) 6
RNA 207
rockwool 190
roentgen, the 169
rotenone 246, 248
Rothampstead Research
      Station 246
[106]ruthenium 177

SAAB company 214
Salmonella sp. 48
sampling
   atmospheric 101
   choice of sites 101
   for GLC 71
   of contaminated land 103
   of plants 104
   of soil 102–3
   of surface water 101
   procedure 143
San Francisco, pollution 26
sarin 249
Scripps Institute 124
sea water, metal levels 144
seal populations 234
sedimentary rocks 143
selective adsorption of metals
      153
selenium, mammalian LD_{50}
      157
Séveso incident 239–42
sewage 267–70
   discharge at sea 267
   disposal 12
   domestic 269

odours 256
overflow 278
sludge 269
   cadmium content 160, 161
   metal content 147
shale oils 7
sheep, poisoning 161
sick office syndrome 191
Silent Spring 7
silicosis 190–1
silyl ethers 63
simazine 254
smelters
   and vegetation damage 134
   arsenic from 148
   as pollution source 18
   at Swansea 149
   lead from 163
   of beryllium 16
   SO_2 from 146
smog
   and tree damage 136
   hydrocarbons in 72
   in London 131
   lachrymatory 203
   toluene in 216
smoke 196–9
   cigarette 190
   maximum concentrations 198
smoking
   and nicotine 248
   risk from 185, 209
soap, usage in USA 218
sodium, in nuclear reactors 175
sodium sulphate, in detergents
      219
soil
   acidification 132
   colloids 37–8
   contaminant assessment 52
   critical contamination 156–7
   organisms 36
   PCB pollution 234
   pollutants 35–42
   porosity 32
   solutions 35
   standards for metals 158
   treatment of polluted 273
solar energy 119, 120

solvent abuse 211, 215, 216, 217
solvents, leakage 274, 275
soot, organochlorine content 235
Spain, SO_2 and NO_x production 137
sphalerite 146, 164
St Paul's Cathedral, damage to 137
straw burning 197
[99]strontium 177–8
structure–activity relations
    of nerve poisons 249–50
    of pyrethrins 245–6
strychnine, lethal dose 238
styrene, USA production 210
sulphonate detergents 218
sulphur compounds
    in petroleum 201
    removal 138
sulphur dioxide see also
        sulphur oxides
    damage to buildings 137
    effect on fish 133
    effect on soils and
        vegetation 133–6
    in lakes 131–3
    toxicity 57, 131
    units of measurement 136
sulphur oxides 129–40 see
        also sulphur dioxide,
        sulphur trioxide
    effect on environment 131–7
    formation and fate 130
    monitoring and control 137–139
    sources 129, 137
sulphur removal 138
    radioactive 171
sulphur trioxide 130
sulphuric acid
    formation 130
    on smuts 196
sun, plasma 186–7
Swansea valley, smelter 149
Sweden, wastes 266
Switzerland
    forest damage 135

wastes 266
synergism 277
    between NO_x and SO_2 136
synthesis
    of aldrin 225
    of carbamates 253
    of chlorophenols 239–40
    of dioxin 236
    of lindane 224
    of parathion 250–1
    of PCBs 232–3
    of PCDFs 236
    of PVC 231–2
    of pyrethrins 246–7
    of vinyl chloride 231
systemic insecticides 251

tailings from mines 145
tanker accidents 275–7
target value for soils 52
TCDD 235–242 see also
        dioxin
    and birth defects 238–9
    at Séveso 239–42
    carcinogenicity 239
    chlorophenols and 235–6
    in air 243
    in humans 240
    in soil 244
    incineration 237
    lethal doses 46
    pollution incidents 237
tea plant 165, 167
temperature inversions 25–6
Tenax, for air sampling 71
teratogenecity
    definition 48
    effect on humans 241, 242
    of mercury 162
    of TCDD 239
tetraethyl pyrophosphate 249
thermal fusion 187
Thiobacillus thiooxidans 146
Thiobacilus ferrooxidans 39
Third World countries
    carbon emissions/capita 118
    energy demand 119
    pesticide use 230–1, 244
    population 10, 118

respiratory diseases 199
thorium, radioactivity 168
threshold levels, of odours 255
threshold values 144
thyroid gland, and iodine 178
tigerfish 229
Tilapia rendalli 229
Times Beach, Mo pollution 237
TLC see also chromatography
    detection 60
    of metal cations 161
    of pesticides 61
tobacco smoke
    amines in 84
    PAH in 209
toluene
    metabolism 44, 45
    reaction in air 205
    toxicity 216
    USA production 210
toxic equivalent factor (TEF) 238
toxicity
    categories, ADME 45
    of pyrethrins 247
toxins, lethal doses, NOAEL 46
trace enrichment
    of chlorophenols 86
    of pesticides 85
trace metals
    and acidity 132
    by coprecipitation 39
    replacement in soils 38
transfer coefficients 154–5
transformer fires 234–5
transport
    as pollution source 19
    in cities 126–7
tree damage, classification 135
trichloroethane
    for dry cleaning 214
    indoors 221
    levels in air 217
    ozone depletion 212, 213
trichloroethylene 74
trichlorophenoxyacetic acid (2,4,5-T) 226, 227, 238

trigger concentrations for risk
    assessment 51–5
tsetse fly 231
tumbleweed 252

United Kingdom (UK)
    forest damage 135
    Forestry Commission 134
    nuclear stations 181
    petrochemicals 202
    SO_2 and NO_x production
        137
    waste production 267
ultraviolet *see* UV
Union Carbide Company 253
United Nations 122, 123
    Environmental Programme 9
units, conversion 136, 280
uranium
    dioxide 173, 174
    experimental toxicity 155
    hexafluoride 173
    in coal 148
    mammalian LD_{50} 157
    radioactivity 168, 173
urban pollution cycle 204–5
United States of America
        (USA)
    adhesives production 211
    Atomic Energy Commission
        181
    chemical production 210
    chlorine production 223
    classification of hazardous
        substances 17–20
    GNP 125
    hazardous waste 271
    National Institute of Health
        238

nuclear stations 180
PAH levels 208
petroleum production 202
Superfund 275
veterans, dioxin exposure
    238–40
waste production 264
UV light 269
    absorption
        by O_2 109–10
        by PAH 81
        by pesticides 85
        by toluamides 83
    and oxidation 277
    in synthesis 223

Van Deemter equation 68
vanadium in petroleum 148
vehicle usage 126–7
ventilation coefficient 26
Venus 120
Verona, Mo pollution 237
Vietnamese War 237, 238
vinyl chloride 231–2
    human toxicity 216
    mass spectrum 73
    USA production 210
VOCs indoors 221
volcanoes 141, 144
Vulcan, the 243

Waldsterben 9, 126, 127
Warren Spring Laboratory
    (UK) 137
waste arisings in UK 267
waste disposal as pollution
    source 18
wastes 263–77

waste, solvents 214
water
    analysis of polluted 85–6
    hardness 152–3
    in vivo reactions 34
    metal levels in 144
    PAH levels in 209
    solvent pollution 215
    standards for metals in 158
    supply in Third World 5
West Germany
    NO_x emissions 126–7
    tree damage 134, 135
wheat bulb fly 229
white rot fungus 245
WHO
    and birth defects 243
    guideline
        for aluminium 164
        for chloroform 216
        for drinking water 57
    smoke limitation 198
wood
    as fuel 11, 12, 117
    burning 197
    composition 200
    preservatives, metal content
        147

X-ray diffraction 189
xenon, as moderator 183
xylene, USA production 210

Yusho incident 233

zeolites 220
zinc ores 145–6
zinc, principal features 163

QMW LIBRARY
(MILE END)